QUANTUM PHYSICS

WHAT EVERYONE NEEDS TO KNOW

QUANTUM PHYSICS

WHAT EVERYONE NEEDS TO KNOW

MICHAEL G. RAYMER

OXFORD
UNIVERSITY PRESS

OXFORD

UNIVERSITY PRESS

Oxford University Press is a department of the University of Oxford. It furthers
the University's objective of excellence in research, scholarship, and education
by publishing worldwide. Oxford is a registered trade mark of Oxford University
Press in the UK and certain other countries.

Published in the United States of America by Oxford University Press
198 Madison Avenue, New York, NY 10016, United States of America.

Library of Congress Cataloging-in-Publication Data
Names: Raymer, Michael G., author.
Title: Quantum physics: what everyone needs to know / Michael G. Raymer.
Description: New York, NY : Oxford University Press, [2017] |
Includes bibliographical references and index.
Identifiers: LCCN 2016050787| ISBN 9780190250713 | ISBN 0190250712
Subjects: LCSH: Quantum theory.
Classification: LCC QC174.12 .R387 2017 | DDC 530.12—dc23
LC record available at https://lccn.loc.gov/2016050787

1 3 5 7 9 8 6 4 2

Paperback printed by LSC Communications, United States of America
Hardback printed by Bridgeport National Bindery, Inc., United States of America

To Kathie

CONTENTS

2. Quantum Measurement and Its Consequences   22

3. Application: Quantum Data Encryption   54

4. Quantum Behavior and Its Description   70

7. Milestones and a Fork in the Road 131

8. Bell-Tests and the End of Local Realism 138

9. Quantum Entanglement and Teleportation   171

12. Application: Sensing Time, Motion, and Gravity with Quantum Technology   **231**

13. Quantum Fields and Their Excitations   **253**

PREFACE

This book explains to nonscientists what quantum physics is all about and the new technologies currently being developed based on the uniquely quantum behaviors of light and matter. It aims to answer in this context, the question, "What does everyone need to know about quantum physics?" My view, as a scientist working in quantum physics, is that a book for *everyone* shouldn't try to explain *everything*. Quantum physics is far too broad a subject. Instead, I asked myself, what is the *one* idea that is most significant about quantum physics that everyone can understand? To many physicists, the most significant aspect of quantum physics is the way it causes us to reexamine our deepest underlying concepts of the physical world—that is, our scientific worldview.

Quantum physics recognizes that the world behaves probabilistically, and does so in a way that clashes with our everyday perceptions. This idea contains two main points. First, the physical world does not behave like clockwork; knowing the current state of affairs as perfectly as possible does not ensure we can predict precisely the future state of affairs. Future events are not predetermined. We can talk meaningfully only about more or less probable futures. Likewise, observing the current state of affairs as perfectly as possible does not tell us exactly the previous state of affairs. While it might not be surprising that it's impossible to predict exactly the behavior of complex systems such as a human, the study of quantum physics shows this is true even for the simplest possible objects, such as a single electron.

Second, consider that when we observe a previously hidden object for the first time, our everyday perceptions are that its appearance and behavior at that moment are direct indications of how they were just before observing them. If you open a gift package and observe a green jewel, you naturally assume it was green before you opened the package. Such an intuitive worldview breaks down at the quantum level when one observes the properties and behaviors of elementary objects such as electrons.

Although you might at first think it is not so odd that intuition breaks down when observing the properties and behaviors of such small objects, most physicists believe the more deeply you think about this fact, the odder it becomes. Nobel Prize–winning physicist Murray Gell-Mann said:

> The discovery of quantum mechanics is one of the greatest achievements of the human race, but it is also one of the most difficult for the human mind to grasp. . . . It violates our intuition—or rather, our intuition has been built up in a way that ignores quantum-mechanical behavior. (Gell-Mann, *The Quark and the Jaguar: Adventures in the Simple and the Complex* [New York: W.H. Freeman, 1994, 123])

A main goal of this book is to help you think about and understand the situation deeply enough so you can appreciate just how counterintuitive the quantum physics worldview *is*, which has been confirmed time and time again by laboratory experiments. In fact, in 2015, as I was writing this book, separate experiments were carried out in three different laboratories providing the strongest evidence ever that the world in which we live is truly incompatible with any explanation based on an intuitive, classical physics worldview. The classical physics worldview, which now seems to be untenable, presumes that each elementary object has definite properties and predetermined behaviors independent of how or if they are observed. *Science* magazine named this accomplishment as one of the Top Ten results in all of science for the year. As a result of these and many previous experiments, which are explored in this book, our scientific worldview has changed forever.

A changing scientific worldview often goes hand in hand with a technological revolution, with the two reinforcing each other in a discovery–innovation feedback loop: agriculture and the advent of human specialization, the printing press and the spread of literacy, eighteenth-century physics and the industrialization of production, the computer and the advent of online social media, and so on. Today, after one hundred years of studying quantum physics, a new breed of technology is arising: quantum technology. The new technology is birthed from a newfound understanding of how the world works. Although it works in counterintuitive ways, scientists and engineers can harness these quantum phenomena to accomplish new kinds of tasks: unbreakable message encryption, ultraprecise sensors of gravity and of acceleration, computers that can compute answers to problems exponentially faster than any computer built to date.

My goal is that a curious, persistent reader with no background in physics will be able to follow the evidence and logic used to explain why we believe quantum theory to be correct, as far as it goes, and how it can be put to good practical uses.

ACKNOWLEDGMENTS

Most important, I thank Kathie Lindlan, my wife, who lovingly encouraged and supported the writing of this book. I also thank my father, Gordon Raymer, for reading the manuscript and offering excellent technical and stylistic suggestions.

I want to acknowledge and thank those others who were most instrumental in the creation of this book. I thank physicists Mark Beck, Steven van Enk, and Dennis Hall for constructive comments on portions of the text. And, besides the many other physics friends and colleagues from whom I have learned over the years (you know who you are), I especially thank two young physicists: Chris Jackson and Dileep Reddy. Together, we developed and taught a course at the University of Oregon on quantum physics for students who were not intending to become scientists (although some decided to be!). The goal of the course was to distill the big ideas of the subject into concepts that can be described without using mathematics. In addition, Chris commented on some early drafts of this book and aided me in developing some of the language used to express the concepts of quantum science to a general audience. I'm grateful for Chris's and Dileep's helpful insights and enthusiasm. Of course, if there are any errors in the presentation, I claim them as my own.

I thank Oxford University Press editor Jeremy Lewis for inviting me to author the book and for shepherding it through the process. And big thanks go to copyeditor Catherine (Cat) Ohala, who provided the highest level of language editing and feedback on parts of the text.

A NOTE FOR EXPERTS ABOUT LANGUAGE

Because this book is for a general audience, I have made an effort to use as ordinary-sounding language as possible to describe technical terms and concepts, departing in some cases from standard physics terminology. Thus, 'quantum-state amplitude' becomes 'possibility,' 'state vector' becomes 'state arrow,' 'wavelength' becomes 'full-cycle length,' 'period' becomes 'full-cycle time,' and so on. I made similar replacements of technical symbols by more familiar-looking ones: 'ket,' $| \, \rangle$, becomes 'parentheses,' ( ); 'direct product,' $\otimes$, becomes 'and' or 'ampersand,' &; and so on.

QUANTUM PHYSICS
WHAT EVERYONE NEEDS TO KNOW

1

QUANTUM PHYSICS

What is quantum physics?

Quantum physics is the study of matter and energy—the basic constituents of the physical world—in the 'quantum realm.' The quantum realm encompasses those aspects of Nature that cannot be explained using classical physics. By 'classical physics' we physicists mean the theory of Nature devised from the 1600s onward by Isaac Newton and others, who built their theories based on the behavior of familiar objects such as rocks, planets, oceans, clouds, wheels, gears, pulleys, clocks, and steam engines. Because of the mechanical nature of many of these things, the theory of classical physics is also called 'classical mechanics.' The theory was expanded during the 1800s to encompass electricity and magnetism, which are more difficult to visualize, but in those days they were also explained in more or less mechanical terms using the basic concepts of classical physics.

Thus, the classical physics theory of Nature was concerned largely with so-called particles (discrete bits of matter moving through space and time) and force fields (influences that establish forces between objects that are not in direct physical contact). For example, electric and magnetic fields establish forces between electrically charged objects and lead to phenomena such as radio signals and light waves that exist over regions much larger than the size of single particles.

Initially, around 1900, when scientists were first figuring out the makeup and structure of atoms, they naturally perceived that electrons, protons, and neutrons had to be particles, and that their behaviors would be well described by classical mechanics. They imagined electrons as being like tiny planets orbiting a larger atomic nucleus playing the Sun's role. But to their shock, when they 'ran the calculations' using Newton's theory of classical mechanics and those of electromagnetism, they found that the predictions of the theory were completely wrong when compared with the results of real-world experiments!

This historic situation drove an intellectual revolution between 1900 and 1925, which in many ways had as great an effect on humanity as did, for example, the French and American political revolutions just more than a century earlier. Classical mechanics was supplemented by the far more powerful theory called 'quantum mechanics,' or simply 'quantum theory.' I say *supplemented* rather than *overthrown* because classical mechanics is still an extremely useful theory, which yields highly accurate predictions for phenomena on the human-size scale. We don't need to use quantum mechanics (although we could) to describe the motions of planes, trains, and automobiles, for example. But we do need 'quantum' mechanics to gain an understanding of the working of electrons and other atomic-scale phenomena.

The challenge for us is that quantum mechanics is a highly abstract theory, making it hard to fathom its true meaning. The good news is that a straightforward use of the quantum theory yields extraordinarily accurate predictions for every phenomenon to which it has been applied. For example, using the ideas of quantum theory, physicists were able to understand how electrons travel through pieces of semiconductor crystals that now make up most of today's electronic devices. Without such an understanding, engineers could never have invented modern computers, which now power the Internet and thus the information society.

The big questions this small book attempts to answer are: What aspects and behaviors of atomic-scale objects and force fields related to them cannot be described using classical physics? How are we to understand these behaviors using quantum theory? And to what good uses can this knowledge be applied? The latter question leads us to explore some very interesting and recent applications of quantum physics in a new field of research and development called quantum technology.

How does quantum physics affect everyday life?

An understanding of quantum physics enabled the invention of many familiar technologies: the laser, the light-emitting diode (LED), the transistor, semiconductor-based electronics including computers and smartphones, high-capacity magnetic disk drives for computer data storage, all-electronic memory used in flash drives and laptop computers, and liquid-crystal displays (LCDs) that are used in nearly all information technology devices. A less familiar invention that emerged recently from quantum physics research is highly secure data encryption. This invention is all the more important now, with recent revelations about the difficulty of protecting information, and the degree to which interceptions of data traffic on the Internet are attempted by unintended persons or agencies.

Modern electronics, including computers and smartphones, rely on the quantum physics of electrons. Lasers, which appear in a wide range of technology and consumer products, create light using the quantum nature of photons. You might wonder: What *are* electrons and photons, and how do they behave? How do physicists explain the seemingly strange behaviors of electrons and photons using quantum theory? What does the word 'quantum' really mean?

One might also be curious about the many news reports touting this or that breakthrough in so-called quantum computing or quantum technology. You might wonder: Why does the word 'quantum' so excite some technologists? What can quantum technology do for us that classical technology can't? Could new breakthroughs lead to creating the technology of the future? Answers to these questions are explored in this book.

What is a physics theory and what is the program of physics?

A physics theory is a way of reasoning made up of a set of well-substantiated concepts or principles that we use to construct 'models,' which are conceptual representations of natural phenomena. A good physics theory captures or encapsulates many features and behaviors of some broad class of physical systems. It compresses a general description of a large portion of Nature into concise statements or principles. Almost always such a compressed description is expressed using mathematics. From this viewpoint, physics is a human endeavor to construct mathematical models of the physical world. To be considered an established scientific theory, it must first survive rigorous experimental testing, during which researchers try to find situations in which it might fail.

If a physics theory passes all the tests to which it is subject, then it may be thought to be correct, and then it can be used reliably to create models of particular situations. But note that scientists can never really *prove* a theory is absolutely correct—only that it works in all cases tested so far. There is always the chance the theory can be superseded by a better, more complete, theory. On the other hand, it is possible to disprove a theory if experimental observations go directly against it.

Physics theories can do more than simply predict what will happen in a given situation. Ideally, they explain, through their many interlinked details, how a phenomenon happens and, in some sense, why it happens. But, to be honest, when pushed to the limits of our fundamental knowledge of Nature, currently the only answer that physics can offer for "Why?"

is, "That's the way it is." We have learned 'the way it is' by experimentation.

So, what is the 'program' of physics? That is, what are physicists striving to achieve? Why do humans want to develop mathematical models of the physical world? There are two main reasons: curiosity and utility. All physics discoveries, although most are driven initially by curiosity, have the potential for useful application. In some cases, the time lag is longer than in others. For example, the physics discoveries leading to the transistor led immediately to useful microcircuitry, which began the current computer revolution. On the other hand, Einstein's discovery of the general theory of relativity in 1915 was not applied practically until about eighty years later, when that theory was built into the global positioning system (GPS), which has revolutionized many aspects of our lives.

Why do we use the word 'model' when referring to physics?

This question gets to the heart of the purpose and role of science. Long ago philosophers believed that natural philosophy, as they then called science, could reveal the true nature of things in the world. In modern times, a different viewpoint generally prevails. A common view of science now is that it provides conceptual models of the behavior of the world, rather than revealing its true underlying Nature (what it "really is").

In science, a 'model' is a mental or conceptual construct used to represent what goes on in the real world. The model is designed to perform in such a way that we can predict how the item being modeled actually performs. Such models are usually described mathematically. An example of a model is a computer program that climate scientists use to make their best predictions of the effects of adding carbon dioxide to Earth's atmosphere. It is important to distinguish between a conceptual model and the system the model represents. By analogy, a toy train might be an excellent model of a real train, but no one would confuse the toy model for the real thing.

Quantum physics is an attempt to model Nature at its most fundamental level, but we should not confuse the quantum physics model (that is, the collection of concepts and mathematical representations) with the real thing (Nature). This kind of thinking, if taken too seriously, can lead to "fairytale physics," in the words of science author Jim Baggott.[1]

Quantum physics had a lot to do with the historical change to the viewpoint that science provides conceptual models only. Because we cannot see, or even infer, what electrons really are, we are forced to work at a more removed, more abstract level when talking about Nature at the quantum level. And because all things are made of 'quantum stuff,' many scientists believe the same insight holds ultimately for everything.

Why was 2015 an especially good year for quantum physics?

As I worked on writing this book, three groups of scientists announced successful experiments verifying for the first time that classical physics theory cannot explain observed measurements on a pair of separated objects that were prepared to have correlated properties. Physicists in Delft, Netherlands; Boulder, Colorado, United States; and Vienna, Austria, carried out measurements on distant but correlated objects that put an end, once and for all, to the classical worldview called 'Local Realism.' This worldview is based on the assumptions that physical objects carry with them definite properties or 'instructions' for how to respond to a measurement being performed on them, and that physical influences acting on any object cannot travel faster than the speed of light. In the classical worldview, two objects can have correlated properties; for example, two balls can be prepared to have the same color although the actual color is unknown. The balls' colors are fixed before they are observed, and if one ball's color is observed, the other's is known immediately as well.

Experiments meant to test Local Realism are called 'Bell-tests' after John Bell, who first proposed such experiments. Local Realism as an assumed basis for physical theory has now been proved false by such experiments. This remarkable conclusion is based on the fact that measurements on two distant objects can yield random outcomes with no fixed, preordained values, yet can—at the same time—display remarkably well-ordered coordination between the distant outcomes. This result flies in the face of commonsense ideas of how the world works. (Again, in the classical view, results of observations may appear to be random, but they are fixed before the measurement is actually carried out.)

On the other hand, quantum theory is perfectly capable of modeling and, in a sense, explaining these experiments without appealing to the concept of fixed, preordained measurement outcomes. This means that quantum theory is inconsistent with Local Realism, as was first proved theoretically by John Bell during the 1960s. These facts seem to have deep philosophical implications about the nature of reality. It remains a mystery how such strong correlations can occur at all when the outcomes of distant experiments cannot be thought of as revealing predetermined values of the quantities being measured.

Chapters of this book are devoted to explaining the Bell-test experiments and how so-called quantum entanglement explains the results.

Why are some objects well described by classical physics models whereas others require a quantum physics description?

There are two main reasons: smallness and coherence, each of which is summarized briefly here. Smallness can refer to different aspects of objects: smallness of size or smallness of energy content. If the object is roughly the size of an atom (about 10^{-10} meters), then it almost certainly cannot be modeled accurately using classical mechanics, and it must be described by the more accurate quantum theory. But, interestingly, the opposite is not necessarily true; objects as large as a millimeter (about a twentieth of an inch) have been observed in experiments displaying behaviors that indicate a quantum nature.[2]

Smallness (or lowness) of energy content could, for example, refer to a tiny electric current in a metal wire (a superconductor) at a temperature only slightly greater than absolute zero (–273 degrees Celsius or –459 degrees Fahrenheit). Low temperature means small or low energy. Or it could refer to a feeble flash of light containing only a tiny fraction (say 10^{-21}) of the energy emitted by a one-hundred-watt bulb in one second. Such a flash of light is said to contain just one 'photon' of light, which refers to the smallest discrete amount of energy that light of a certain color can carry.

A discrete amount of energy such as this is also called a 'light quantum.' The plural of quantum is 'quanta.' Therefore, for example, a burst of light with a large amount of energy is said to contain many quanta. This discreteness of the energy carried in light, which we explore in more detail later, is the origin of the name 'quantum physics'.

In principle, a single quantum entity such as a photon could extend across a very large volume—for example, many kilometers. Although such a photon would be large in size, it would be very small, or low, in energy content, and so the quantum theory would still apply to it.

The second general reason an object may require a quantum description is 'quantum coherence.' Quantum coherence is a subtle concept and it cannot be understood properly until after one understands how the state of an object is described using quantum theory. To give you a flavor of what is to come in later chapters, quantum objects can behave in ways that appear random, although there is no obvious underlying physical cause of this randomness. For the case of an electron, quantum coherence enters the theory in how it accounts for the different possibilities that may exist before the electron's location is observed. In a sense, the usual rules of logical thinking, such as saying, "It is located here or it is not located here," do not apply to quantum objects. Instead, it is said, "Both possibilities must be superimposed in our thinking and not considered separately." Quantum coherence makes such a *superposition of possibilities* physically realizable, as explained in later chapters.

What are the elementary entities that make up the physical universe?

This is a big question, and answering it has been the aim of physics research for centuries. The simplest answer is that nearly all matter we can observe directly with our simple human senses is made up of atoms, which are comprised of electrons, protons, and neutrons, as stated previously. Electrons are thought to be 'elementary' constituents of matter in the sense that they are not made of yet-smaller constituents. (Notice I am *not* using the word 'particle' here, to avoid any misleading impressions that word might convey.) On the other hand, protons and neutrons are comprised of smaller elementary constituents called quarks, which have the curious property that they cannot exist on their own outside of the groupings of quarks that make up objects such as protons or neutrons. Their existence is known through experiments begun during the 1960s in which fast-moving electrons were aimed at protons, and the pattern of the deflected electrons indicated that protons have an internal substructure. A detailed model based on quantum physics was developed in which each proton or neutron is composed of three quarks of specific types. The model also made concrete predictions about further experiments, all of which have been verified, so we have good reason to believe the quark model is correct.

Another important entity is the electromagnetic field, which refers to the combination of the electric fields and magnetic fields that surround electrically charged objects or magnets. These 'fields of influence' not only transmit static electric forces and magnetic forces, they also make up radio waves and light waves, as mentioned earlier. Light and radio waves carry energy. Energy is defined most simply as the capability to cause motion. For example, a radio wave impinging on a radio antenna causes electrons in the metal of the antenna to move (the motion of which can be detected and amplified to drive audio speakers). These phenomena are well described by classical mechanics.

At the quantum level, light can be viewed as being comprised of photons, which can be thought of very roughly as particlelike entities that carry the energy in a radio wave or light beam. Photons are elementary in that they are not comprised of constituents. It turns out that, for a photon, there is no clear concept of a precise 'position' or 'location'—concepts we associate with particles. Yet, as we will see, they do behave in certain ways that we expect particles to behave. At the same time, we know that light has some wavelike behavior, so photons must also somehow carry wavelike behaviors. Therefore, a photon is neither a particle nor a wave in the classical sense.

This verbal dance I am doing to try to describe photons illustrates the difficulty of saying what a photon "really is," and the difficulty of visualizing accurately how a photon behaves. Physicists have gotten used to this ambiguity and have no trouble deploying the mathematical machinery that we use to predict the outcomes of events involving photons. Yet even physicists have a hard time picturing in a simple way how all this "really happens." For some, like me, this puzzlement makes quantum physics all the more intriguing and fascinating.

There are other kinds of particlelike entities, with exotic names such as mesons, muons, positrons, and neutrinos. And there are fields other than the electromagnetic field—for example, the strong force, which is a field responsible for holding protons and neutrons together in an atomic nucleus. A rather exhaustive theoretical model, based on quantum physics and called the Standard Model of particle physics, encompasses all the known entities mentioned here, plus others I won't mention. This model, which is mathematical and highly abstract, predicts successfully essentially all the known processes involving all the identified elementary entities in Nature. The capstone of discovery that supports the Standard Model most strongly was the detection of the Higgs boson in 2012.

The invention of the Standard Model, and its experimental confirmation, are together an exceptional achievement for humanity. Yet, there are still large unknowns in the universe: so-called dark matter and dark energy, the existence of which astronomers infer from analysis of the motions of distant galaxies. In fact, it is estimated that about ninety-five percent of the universe is comprised of these as-yet-unknown entities. When they are identified, it is expected the Standard Model will need to be updated. Even so, it seems likely to many physicists that the basic way in which quantum theory models the world will remain intact.

How is light different in classical and quantum descriptions?

Light is familiar to everyone, and it was the first phenomenon to be described by quantum theory, so let's choose it as our first example for a detailed discussion. As mentioned earlier, light is an electromagnetic wave that carries energy. In the classical theory, light was conceived as having its energy spread smoothly throughout the region that a light beam occupies. For example, when a laser pointer is aimed at a screen, the energy of the light is spread smoothly within the beam between the pointer and the screen, and spread smoothly across the area of the screen being illuminated. This is in analogy with water waves created by a boat on a lake; the energy of the waves is spread continuously throughout a region of the surrounding water and the waves arrive at the lakeshore in a smooth, spread-out way.

Any wavelike motion has an associated 'frequency.' The frequency of light refers to how rapidly the electric and magnetic fields are vibrating or oscillating in the light wave. Frequency is related directly to color; blue light has a greater frequency than red light.

This description of light carries over quite well to the more accurate quantum description, with one main exception: Although the energy in a light wave is indeed spread out, when it is extracted from the light wave, the process appears to occur in small lumps. We call this behavior the 'discreteness of light detection.'

What are consequences of the discreteness of light detection?

When you look at an object being illuminated brightly by a light bulb, you perceive the object's shape, color, and texture. If it has smooth texture and uniform color, you perceive a uniform surface with uniform brightness. If the light bulb is on a dimmer controller, you can turn up the brightness ever so slightly and perceive a slightly brighter surface. We call such smoothly changing behavior 'continuous.' It is the opposite of 'discrete,' which means that something occurs in steps or comes in lumps. For example, a wheelchair ramp is continuous, whereas a staircase is discrete. As another example, an oil painting is continuous (at least down to a scale not smaller than individual paint grains), whereas a digital photograph of the same painting when viewed on the camera's display is discrete at the scale of the screen's LCD pixels (which are much larger than a paint grain).

Consider how the situation changes if the brightness of the light source is dimmed drastically. As you might know (and all photographers do), the image becomes grainy. You can see this effect by taking a photo in a fairly dark room and using image-enhancing software to brighten it. You would notice the image does not brighten uniformly; instead, some pixels get much brighter than they "should" be, whereas other pixels remain too dark. A similar effect happens using either digital photo technology or old-style chemical photography film. The word 'grainy' derives from the tiny grains of silver chloride crystals—the light-sensitive components of the film.

The camera's pixel array is almost perfectly uniform. That is, all the light-sensitive pixels are of the same size and have nearly identical responses to light. Despite the array's uniformity, an underexposed array results in a grainy image, because each pixel needs a certain minimum amount of light energy to fall onto it before it can send an electronic signal to the camera's memory. If the total amount of light energy falling on the whole array is very small, then only a small fraction of all the pixels are able to create such a signal, and so the image looks grainy. We might try to explain this graininess by speculating that the light from the bulb is not uniformly bright at each pixel, so naturally some pixels receive more exposure than others.

To make the situation more precise, let's replace the light bulb with a laser that emits a wide beam illuminating the camera pixels. The beam from a laser can be made perfectly uniform across all pixels, and in the classical physics model this means a perfectly steady flow of energy arriving at each pixel. The pixels can be built so that all the light power hitting the array is absorbed. Consider exposing the array to very weak laser light for only one second. In this case we would expect, using classical physics, that below a certain power level no pixels would receive enough energy in the one-second interval to 'fire' and send a signal to the memory. It turns out, however, that this is *not* what is observed. In fact, for very low light brightness, we observe that a few pixels fire and the remaining ones do not. Although the light hitting the array is perfectly uniform, when energy is extracted from it, the extraction process seems to be discrete.

The correct way to understand these results is to follow the quantum physics model, which says that although the electromagnetic field is uniform across the array, when energy is extracted from the field it occurs in lumps, also called 'quanta.' The strange-seeming aspect of this explanation is that it implies energy in the field can be concentrated into one pixel and cause it to fire, although the classical field description does not ascribe enough light power at any single pixel to fire it. It is common to say that this behavior implies some 'particlelike' behavior of the light field when it is detected by the array. This does not mean that light *is* composed of small particles called photons; rather, in some cases, light behaves *as if* it were composed of photons. Albert Einstein received his Nobel Prize for figuring this out in a slightly different context.

Is it possible to create and detect exactly one photon?

Yes, it is, and there are various ways to do it. A conceptually simple way is the following: Isolate and hold just one atom in place—a sodium atom will do nicely—and send a short burst of orange laser light at it. Although the light burst has an unknown amount of photons in it, it can be arranged so the atom absorbs exactly one photon's worth of energy from the laser light. After a brief time elapses, the atom releases, or emits, that energy as a single photon of light, which can be sent in a desired direction using lenses and mirrors.

We can verify by experiments that only one photon is present; send the light onto a half-silvered mirror, which is a piece of clear glass with a thin layer of silver on one side. Roughly half the light power striking such a mirror passes through the mirror, and half is reflected from the mirror to a different direction. This is why, when you wear silvered sunglasses, your view is dimmed, but you can still see pretty clearly. Each of the detectors, shown as cylinders in FIGURE 1.1, potentially receives some light energy. However, it is always observed that only one of the two detectors fires on a given trial of this experiment. This is easily understood when adopting the oversimplified viewpoint that light behaves in some ways similar to a stream of particles, each of which either passes through the mirror or is reflected. If only one particle were present, only a single detector could fire.

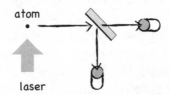

Figure 1.1 A burst of light from a laser puts energy into a single atom, which then releases that energy in the form of light. The light emitted passes through a half-silvered mirror and potentially strikes two light detectors. During each trial, one or the other detector fires—but not both.

But it is not quite correct to say a particle of light arrived at the mirror and randomly "made a choice" to travel toward one or the other detector. We know this is not quite right, because it's possible to replace the detectors with other devices that lead us to a conclusion that the light arriving at the mirror behaves more wavelike than particlelike. We discuss these kinds of experiments in later chapters. Another way to say this using quantum language is that the light's electromagnetic field, which behaves like a wave, arrives at both detectors, but it contains enough energy to fire only one of them. The one-photon's worth of energy is somehow collected in one of the detectors. So, although we can say that only one photon was present in this experiment, we cannot say the photon "is a particle." It's more accurate to say that the light contains one-photon's worth of energy, and that when being detected this energy is indivisible. In this sense, we say that a photon is elementary; it is not broken into subparts.

How was quantum physics discovered?

The detailed story of how quantum physics was discovered is fascinating, and many books tell it. But with hindsight, it seems to me that the great struggle to invent quantum theory during the early twentieth century says more about the difficulty humans had (and still have) in going beyond classical physics thinking than it says about the facts of quantum physics. So this book does not dwell on the historical aspects of physics. In this book, historical details are given when they add to the clarity of the physics being discussed. Here I give a thumbnail sketch of historical highlights, and how each added to the growing body of knowledge about quantum physics. At the same time, I introduce some quantum physics concepts not mentioned previously—in particular, the concept of quantum fields.

Early philosophers and scientists such as Newton had questions about the nature of light: Is it waves or particles or neither? But it was not until 1900 that solid scientific evidence was gathered that began to answer the question. The story goes as follows: When a body of normally black material is heated to high temperature, it gives off light of different colors, much like a glowing metal burner on a cook stove. If the object is hot enough, the light looks whitish. The spectrum of light can be separated using a prism and the brightness of each color can be measured with a light detector. When these color-dependent brightnesses were compared with the predictions of classical physics theory, that theory was found to be faulty.

German physicist Max Planck discovered that the problem with the classical theory lay in the seemingly reasonable assumption that energy could be exchanged between the hot material and the light in any amount within a continuous range of energies. In an attempt to get a better agreement between the experiment and the theory, Planck tried changing just one aspect of the theoretical model. He made the new assumption—radical for his time—that the possible energies exchanged between the material and the light were not continuous, but discrete; that is, they occur in steps, as in a staircase. He posited that the size of these energy steps is proportional to the frequency associated with the color of the light being considered. The constant of proportionality is now known as Planck's constant. To the astonishment of the physics world, this revised model when solved mathematically was in perfect agreement with the experimental measurements of the different colors' brightnesses.

Albert Einstein was inspired by Planck's success to suggest a general hypothesis concerning light. He supposed that light of a given color could only have energy content that is discrete, not continuous as would be expected from classical physics. He called these discrete amounts of energy by the name 'light quanta.' And he called a single one of these a 'light quantum.' We now call them photons. Einstein further hypothesized that light quanta are indivisible; that is, they interact with light-absorbing materials as wholes. Each photon is either absorbed or it is not; it can't be partially absorbed. He developed these ideas into a working theory that predicted correctly how atoms absorb and emit light. His equations turned out, fifty-five years later, to be the theoretical impetus leading to the discovery in 1960 of the laser. This is another illustration of the fact that basic-science discoveries are nearly always at the root of the most important technological inventions, although the time lag is sometimes quite long.

Not long after Planck pondered the smooth rainbow-like spectrum of light emitted by hot objects, other scientists were studying the light given off by a gas or vapor containing atoms of only one element (say, neon) when an electric current is passed through it. This is the light we see every day from fluorescent light bulbs. At the time, it was understood that an atom of neon is made of a nucleus containing ten protons and usually ten neutrons, surrounded by ten electrons. And it was known that electrons are 'matter,' in that they have mass, unlike photons, which have zero mass. It was assumed that electrons behaved like tiny planets orbiting the nucleus, as if it were a tiny sun. For this model of the atom, classical physics theory predicted that when a neon atom with some excess energy held by its electrons gives up some of this energy, light of any color within a continuous range could be given off. But experimenters noticed that the light actually given off consists of only a few well-defined colors, not the expected smooth rainbow of colors. This was a great mystery because classical physics theory could not account for this observation.

To make a long story very short, by 1925 it had been realized that the flaw in the earlier theory was in assuming that electrons behaved like tiny planets. That is, electrons should not be viewed as being tiny particles, or bits of matter that follow definite paths around the nucleus. Louis de Broglie—then a physics graduate student at the University of Paris—first postulated that perhaps electrons behave in some ways like waves, which was a very nonparticlelike view of electrons. Erwin Schrödinger was able to codify this viewpoint into a mathematical theory, which was able to predict perfectly all the possible wavelike patterns that an electron in a given type of atom could make. He proved that each wavelike pattern is associated with a particular wave frequency, and therefore, following Planck's idea, with a particular energy. He further realized that when an electron changes from a higher energy pattern to a lower energy one, it gives off light of a particular frequency, and thus a particular color. Schrödinger's equation was able to predict correctly all the discrete color patterns observed in experiments on atomic-gas light bulbs. Because this was such a momentous discovery, and because Schrödinger's equation is beautiful and inspiring to physicists, I reproduce it in the Notes at the end of this chapter.[3]

Do electromagnetic fields have a quantum nature?

Again skipping over many important historical details, we come to the point in 1925 when Max Born, Werner Heisenberg and Pasqual Jordan wondered about the following: Physicists believe the electric and magnetic fields are real 'things' in the world, well described by the classical theory called Maxwell's equations; but Einstein had previously muddled the nice classical theoretical picture of these fields by introducing the fairly vague notion of light quanta.

Born, Heisenberg, and Jordan wanted to know: What is the relation between the field concept and the photon concept? They thought that if electromagnetic fields are real 'things,' then they, too, should be able to be described using quantum theory. They developed a mathematical formalism we call the 'quantum theory of fields' or, for short, 'quantum field theory.' They showed that if one postulated that electromagnetic fields, not photons, are the fundamental 'things' in Nature, then the photons, which express the particlelike nature of light, arise naturally as a consequence of the quantum field theory.

In 1927, Paul Dirac, then a twenty-five-year-old with a freshly minted PhD from Cambridge University in England, was the first to apply the new quantum field theory of light to the question: How do atoms absorb and emit light quanta? He was able to derive, using his new mathematics, the same results that Einstein had postulated earlier following Planck's ideas. Dirac's landmark result put quantum field theory on a firm footing.

Soon thereafter, physicists—starting with Pasqual Jordan and Eugene Wigner—tried applying the same reasoning to electrons, which are a type of matter. Electrons have mass, whereas light does not. They argued that if the theory of light can be built by starting from the viewpoint that the electromagnetic field is the fundamental 'thing,' and photons are simply physical aspects of the quantum light field, then maybe the same reasoning can be applied to matter. Perhaps there is such a thing as an 'electron matter field,' and electrons are simply manifestations of the quantum nature of the matter field. This idea was found to hold up nicely from a mathematical point of view, so physicists started believing that perhaps matter fields are actual physical 'things' in Nature. But, initially, there was no experimental evidence supporting this idea.

Finally, during the 1960s, experiments began showing physical effects that could be best explained by using the theoretical concepts of quantum fields of both kinds: electromagnetic and matter. The experiments involved the direct interactions between these two quantum fields. For example, it was observed that when a very-high-frequency electromagnetic field interacted with the electron matter field, the electromagnetic field lost one photon's worth of energy and the electron field gained one electron's worth of energy. In particlelike language, we say one photon was destroyed and one electron was created. At the same time, a completely new kind of matter field was observed to gain energy, creating a particlelike object called a positron. A positron has all the attributes of an electron except that its electric charge is positive rather than negative. Theorists found that, instead of talking about the appearance and disappearance of particlelike objects, it is more natural to speak of the various fields gaining and losing quanta of energy. This suggests the quantum fields—more so than the particles—are real 'things.' Therefore, through this combination of experiment and theory, most physicists became convinced that quantum fields are the elementary entities that make up the physical universe. In this view, so-called 'particles' are merely physical manifestations of the quantum fields.

The quantum theory of fields eventually evolved into the Standard Model of particle physics, mentioned previously. In this theory, which is the most advanced we currently have, every different kind of elementary particle (electron, quark, neutrino, and so on) is considered to be an aspect of its corresponding quantum field. That is, in addition to the electron matter field, there is a quark matter field and a neutrino matter field, and so on. They interact with each other not directly, but by interacting with intermediary force fields, such as the quantum electromagnetic field, the physical aspects of which we call photons. In fact, in the Standard Model theory, all the known types of matter and the known types of forces are described as quantum fields. In this view, all things, including the atoms that comprise you and the forces that hold you together, are interacting quantum fields. You are a walking collection of interacting quantum fields!

Notes

1 The term "fairytale physics" comes from the book by Jim Baggott, which offers a view of the dangers of taking the mathematics in a theory too literally: *Farewell to Reality: How Modern Physics Has Betrayed the Search for Scientific Truth* (New York, NY: Pegasus, 2014).

2 An example of quantum behavior in large objects has been demonstrated in diamond crystals about one millimeter in size. A dramatic experiment showed that the internal vibrational motion of two such diamonds separated by fifteen centimeters can be prepared in a 'quantum entangled' state, the meaning of which is explained in Chapter 9. An open-source discussion of the experiment is in the article "Diamonds Entangled at Room Temperature," 2011, Centre for Quantum Technologies, (http://www.quantumlah.org/highlight/111202_diamonds).

3 Schrödinger's equation, which you don't need to understand, but simply admire, is

$$ i\hbar \frac{\partial \psi}{\partial t} = -\frac{\hbar^2}{2m} \nabla^2 \psi + V \psi. $$

Looking at this equation, before I learned calculus, was one of my main inspirations for studying quantum physics. I was intrigued by the idea that so few symbols could embody so many complex phenomena.

Further Reading

The Bell-test experiments that put an end to the classical worldview called 'Local Realism' are described in Alain Aspect's "Viewpoint: Closing the Door on Einstein and Bohr's Quantum Debate." *Physics* 8 (2015), 123, http://physics.aps.org/articles/v8/123.

2

QUANTUM MEASUREMENT AND ITS CONSEQUENCES

What is measurement in classical physics?

In the classical realm of human-scale objects such as motorcycles and buildings, it's pretty clear what we mean by 'measuring an object.' If you wish to know the length of a wall in your bedroom, you use a ruler or measuring tape to measure it. Or, if you wish to determine the speed of a motorcycle passing by, you could mark off a 30-foot (or 10-meter) section of the street and use a stopwatch to measure the time it takes to travel the distance between marks. The speed is the distance divided by the elapsed time.

There are three aspects of the classical concept of measurement. First, it is presumed we can increase the precision of such measurements without limit if we have the needed instrument for our use. Second, it's clear that making such measurements can be done in a way that does not significantly change the objects being measured. Third, it seems obvious that the length of the wall and the speed of the motorcycle actually *have* values when they are measured and before you measure them. That is, we presume objects inherently possess definite values for the features to be measured before actually carrying out the measurement.

It turns out that at the atomic scale, for quantum objects such as electrons, these aspects of measurement do not hold up as they do in the 'classical' world. It is not obvious that measurements on such an object can be done without changing the object significantly. It's not even obvious that the feature or property of the object you wish to measure in the quantum world actually *has* a definite value before you measure it.

Before trying to explain how these radical-sounding ideas could really be the way the subatomic world is, let's take a tour through the world of polarized light. This provides a nice illustration of the way 'measurement' works in the quantum realm. Along the way, we will see where the concept of 'quantum' enters.

What is light polarization?

Every beam of light has a feature called *polarization*, and it can be put to good use. Polarized sunglasses reduce the glare you would see from the surface of a lake or the hood of a car. The three-dimensional glasses you wear at movie theaters make use of light polarization. And the LCD screen of your computer or smartphone works by using polarized light. To confirm this, try rotating a pair of polarized glasses in front of such a screen.

To understand how polarization works, imagine an arrow flying through the air. This arrow has a pair of feathers at its tail. Let's consider that the orientation of the feathers stays constant as the arrow flies. The orientation of the feathers can be thought of as a feature, or property, of the arrow; we can talk about a vertically feathered arrow or a horizontally feathered arrow, or some case in between.

The polarization of a light beam traveling parallel to the ground is analogous to the arrow's feather orientation. It can be oriented either vertically or horizontally, or in between. What is light polarization? I mentioned in Chapter 1 that an *electric field* is an influence that establishes a force between electrically charged objects that are not in direct physical contact. According to classical physics theory, light is made up of traveling, oscillating electric and magnetic fields. The polarization of light corresponds to the orientation in which the electric field is oscillating as the light travels. The direction of the polarization is perpendicular to the direction the light is traveling, just as feathers are perpendicular to the direction an arrow is traveling. The direction of the polarization can be thought of as a feature, or property, of the light; we can talk about a vertically polarized light beam or a horizontally polarized light beam, or a polarization somewhere in between.

How do we determine or measure light polarization?

Consider the arrow again. For simplicity, let's consider only those cases in which the orientation of the feathers is either vertical or horizontal. What if the arrow is flying through a dark room so you can't see the feathers? How can you determine its feather orientation? You could put it through some kind of test and see what happens. Imagine that the head of the arrow passes through a slot between closely spaced horizontal slats in a fence, as in FIGURE 2.1. If it is a horizontally feathered arrow, it passes through the slot unimpeded. But, if the feathers are vertical and stiff, the arrow is not able to pass through the slot.

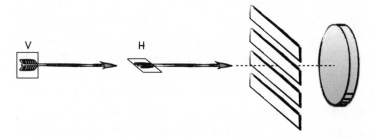

Figure 2.1 Measuring the orientation of the feathers on an arrow. The slots accept horizontally oriented arrows and a target registers their arrival.

If you know beforehand that the arrow's feathers are either vertical or horizontal, you can use this arrangement to determine their orientation. Let the arrow attempt to pass through a slot in the horizontally slotted fence. If you can somehow ascertain that it passed through, you will know without actually seeing it that it was a horizontally feathered arrow. If you ascertain that it doesn't pass through, you know it is a vertically feathered arrow.

To ascertain whether the arrow passes through, you could place a target behind the fence and, with your hand touching the edge of the target, you could feel the arrow's impact if it strikes the target. The target acts to record or 'register' the arrival of the arrow. In physics we would call the fence slot a *selector* and the target a *detector*. Both are needed to accomplish a meaningful measurement.

If a stream of arrows, some with vertical and some with horizontal feather orientations, are aimed toward the slot in the fence, some pass through and some are blocked. This amounts to a classical physics determination or measurement of the feather orientation of each arrow. We can characterize the arrow stream by noting the percentage of arrows measured to be vertical, with the rest being horizontal. For example, we might find a particular stream of arrows was thirty-five percent vertical and sixty-five percent horizontal.

For measuring the polarization of light, we can use a glass window made of a special material through which light of a particular polarization can pass, but the perpendicular polarization is blocked. A piece of such material is called a *polarizer*. If vertically polarized light is incident on such a polarizer that is oriented in a certain way, the light is blocked completely. If that same polarizer is then rotated 90 degrees, then vertically polarized light passes perfectly and horizontally polarized light is blocked. If your eyeglasses are made of that material, then sunlight that has bounced off your car hood before reaching your glasses is mostly blocked, because that reflected light is mostly polarized horizontally.

We can imagine the action of the polarizer to be analogous to the way the slotted fence acts on the arrows, as illustrated in FIGURE 2.2. In this drawing, the short lines show the direction and strength of the electric field at each point where the light is traveling. If the light's polarization is parallel to the polarizer's 'slots,' then the light passes through undimmed; but, if the light is polarized perpendicular to the polarizer's slots, the light is blocked. If a detector that can register arriving light is placed behind the polarizer, then we have a setup that measures the light's polarization, H (for horizontal) or V (for vertical).

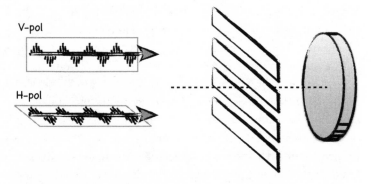

Figure 2.2 Measuring the polarization of light with a polarizer and a light detector. Only horizontally polarized (H-pol) light passes through the polarizer and is registered at the detector. Vertically polarized (V-pol) light is blocked.

To shorten our terminology, let's abbreviate the phrase vertical polarization to V-pol and horizontal polarization to H-pol. We can refer to light being either V-pol or H-pol, or something in between.

What if the light beam contains some combination of V-pol and H-pol light? How do we characterize its polarization? As with the arrow stream, we can do this by determining the fraction of the light beam that is V-polarized or H-polarized. To make this characterization possible, we use a special kind of polarizer that passes H-pol light and, instead of simply blocking V-pol light, deflects it into a different beam. A crystal made of calcite does the trick nicely. Calcite is a common transparent mineral, samples of which can be bought in rock collector shops. It causes double refraction—meaning light beams of different polarizations are bent or refracted differently when entering the crystal.

This effect of light passing through a calcite crystal is illustrated in FIGURE 2.3, which shows the calcite crystal in an orientation we call the horizontal/vertical or (H/V) orientation. The arrow drawn on the crystal indicates in which direction the light with polarization parallel to the arrow gets deflected. We call this arrow the *axis* of the crystal. Light with polarization perpendicular to the crystal axis is not deflected.

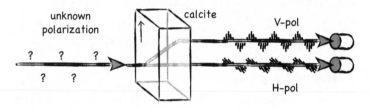

Figure 2.3 Characterizing the polarization of light with a suitably oriented calcite crystal and two light detectors.

We place two light detectors behind the crystal, one in each emerging beam. A certain amount of light power is deflected toward one or the other detector, which registers a voltage depending on how much light energy strikes it. If the incoming light is purely H-pol, only the lower detector registers light striking it. If the incoming light is purely V-pol, only the upper detector registers light striking it.

What happens if the light contains a mixture of polarizations?

Direct sunlight, for example, contains all polarizations—H, V, and all orientations in between—in equal amounts. We call this an equal *mixture* of all polarizations. In such a case, when we characterize the light as in FIGURE 2.3, we would find equal amounts of light registered at the two detectors. In fact, if we were to rotate the calcite crystal and the detectors together to any orientation, we would still register equal amounts of light hitting the two detectors. That is, the polarization of sunlight has no preferred orientation.

In the general case, a mixture can occur with various proportions of different polarizations. To characterize this light, we can register the fraction of all the light detected as V-pol, and the fraction detected as H-pol. This measurement gives us important information, but it is not the whole story, as illustrated by the next question.

What happens if the light is purely polarized other than H or V?

Light can be purely polarized and oriented at any angle from the V orientation. For example, a light beam can be polarized at +45 degrees from the V orientation, in which case we say it is diagonally polarized, or D-pol. Or it can be polarized at −45 degrees from the V orientation, in which case we say it is antidiagonally polarized, or A-pol, as shown in FIGURE 2.4.

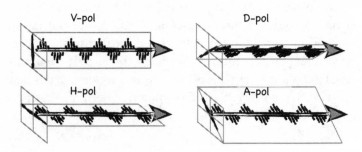

Figure 2.4 Examples of light polarization.

Let's say a light beam is purely D-pol, but you don't know that. You pass the beam through the calcite crystal as in FIGURE 2.3. Can you predict the result? It is found that, when you do this experiment, you register equal amounts of light at each detector. That is, D-pol light somehow contains equal proportions of H and V polarization. In quantum physics, we call such a situation a *superposition of possibilities*. The calcite crystal separates the light into these two possibilities. You might be wondering: We started with D-pol light. How did it change to V and H? This result may seem mysterious, and exploring its underlying cause can lead us to a deep understanding of light and of quantum physics.

Another important fact is that all the light that exited the crystal in the V-pol beam is now V-pol, even though the original beam was D-pol. This fact can be checked by passing the exiting beam through a second H/V-oriented calcite crystal. It is found that all the light in this case exits the second crystal in the V-pol beam, confirming its polarization was V-pol.

We just discussed that purely D-pol light is made up of equal parts of V-pol and H-pol light. Yet it is not merely a mixture of V- and H-pol light, as is sunlight. You can verify this by passing D-pol light through a calcite crystal that is itself oriented at −45 degrees from vertical, as in FIGURE 2.5. As shown, the light would pass through without splitting into two beams and would remain D-pol. You would observe no light in the A-pol beam. In contrast, sunlight, which contains a mixture of polarizations, would split equally into D-pol and A-pol beams, as mentioned previously.

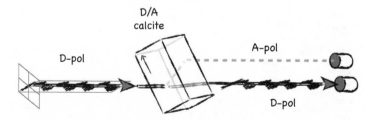

Figure 2.5 Although diagonally polarized light is made of equal parts vertically polarized and horizontally polarized light, when it passes through a D/A calcite crystal (oriented at −45 degrees), it remains D-pol and exits in the corresponding beam.

What is the physics behind these polarization measurement results?

D-pol light is a very particular combination of H-pol and V-pol light. The best way to understand this combination is to recognize that light is a wave moving in (or on) the electric and magnetic fields, as mentioned earlier. The drawings in the previous figures are meant to illustrate this wave. The electric field has a direction associated with it; at a particular location it is aimed, or 'points,' in a direction at a given instant in time. (Recall, the electric field exerts a force on electrons opposite to the direction the field is pointing.) The short lines in the drawings show the direction in which the electric field points at each location. These directions are always perpendicular to the direction the light beam is traveling. In V-pol light, the electric field oscillates rapidly between pointing upward and pointing downward.

To visualize these directions, let's use the analogy of the directions on a map and a directional indicator represented by an arrow, as in FIGURE 2.6. You can visualize the light beam as traveling in the direction into the page—that is, away from you. The directional arrow, which represents the polarization, is shown pointing northeast—that is, in the diagonal direction. (Southwest is called the minus-diagonal direction.) The direction northeast, or NE, is 'composed of' equal north (N) and east (E) components, which combine to make a result pointing NE.

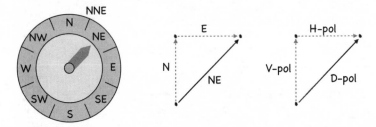

Figure 2.6 Diagonal polarization is composed of equal parts of vertical polarization and horizontal polarization, just as the resultant northeastern direction is composed of equal parts of northern and eastern directions.

Likewise, D polarization is in a direction composed of equal parts of V and H directions. Because the dashed lines shown pointing in the V and H directions can be viewed as the composition of the arrow in the D-pol direction, we call these dashed lines the H and V *components* of the D polarization arrow. A calcite crystal in the H/V orientation does the job of separating the H component from the V component. So, for a D-pol light beam, two beams emerge, with equal powers.

What about the north-northeast (NNE) direction, also shown on the directional dial in FIGURE 2.6? It is tilted more toward the N direction than the E direction, so it is not composed of equal parts of N and E. If you were to draw the triangle illustrating this case, it would have a longer N side and a shorter E side; when combined, they would make up an arrow pointing NNE.

What is coherence and what role does it play?

Recall that sunlight is a mixture of polarizations in all directions. It contains equal parts of V and H components. Pure D-pol light is not such a mixture, although it, too, contains equal parts of V and H directions. The concept of 'coherence' is how we distinguish between these two types of light. Physicists use the name *coherent superposition* to describe a combination of two polarizations that gives rise to a new polarization with a well-defined direction different than that of either component.

The electric field in light oscillates rapidly. FIGURE 2.7 shows a time-lapse series of drawings for D-pol light separated in time by one femtosecond (1 femtosecond [fs] equals 10^{-15} of a second). Initially, at time 0 fs, the electric field arrow points in the diagonal direction D. At 1 fs, the H and V components have shrunk to half their original size, as has the length of the electric field arrow. At 2 fs, all three have shrunk to zero size. At 3 fs, each arrow has grown again, but in directions opposite to the originals. At 4 fs, the arrows have grown to replicate the original pattern but with all arrows in directions opposite to the originals. Here I labeled the components –V and –H. The electric field arrow is labeled –D. Here, '–' is read as 'minus' and means 'in the opposite direction.' If we were to wait another 4 fs, the pattern would revert to its original form that we saw at 0 fs. That is, it takes 8 fs to complete one full cycle of this oscillation, during which the electric field arrow always points somewhere along the diagonal line. An 8-fs full-cycle time identifies this light as infrared light, which is not visible to the eye but is still considered light.

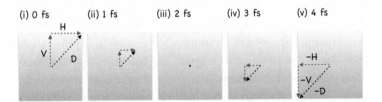

Figure 2.7 (i–v) Time-lapse drawings, from left to right, of a purely diagonally polarized light beam (viewed end-on), showing synchronized oscillations in the strengths of its vertical and horizontal components. Coherence causes the electric field arrow to oscillate in the diagonal direction. The vertically polarized and diagonally polarized components 'cohere' to form the diagonal polarization.

The reason this nicely organized pattern of oscillation takes place is that the V and H components are synchronized. Like synchronized swimmers performing a routine, the two components move in a coordinated way, leading to the motion of the electric field arrow shown. This synchronization of behavior is called *coherence*.

In contrast, if the two components move in a random, chaotic way with no synchronization and thus no coherence, the electric field arrow does not remain pointing along the diagonal line; the polarization is not stable and points 'all over the place.' For example, sunlight does not have coherence of its oscillating V and H components, and so it is a 'mixture' of polarizations rather than a pure polarization.

The example just given illustrates coherence in the context of classical physics theory in which light is viewed as a wave in the electric field. Later in this chapter we explore the idea of coherence in quantum physics after we examine the measurement of photons, which correspond to a quantum aspect of light.

Can we measure the polarization of a single photon?

Recall that in the quantum physics description, light seems to consist of discrete units of energy called photons. A photon is the smallest, or least, amount of energy possible in a light beam of a particular color and polarization. If a V-pol photon enters an H/V-oriented calcite crystal, it exits in the V-pol beam. An entering H-pol photon exits in the H-pol beam.

But what happens if a D-pol photon enters an H/V-oriented calcite crystal, as in FIGURE 2.8? If there are detectors that intercept the outgoing V-pol and H-pol beams, it is observed that the photon is registered at one or the other detector, not both. A photon is an elementary object, and its detection is 'all or nothing.' But, you might ask, which detector registers it? For D-pol light, the answer is that both are equally likely!

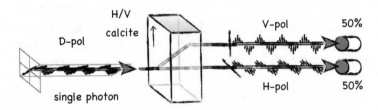

Figure 2.8 A single diagonally polarized photon enters a horizontally/vertically (H/V) oriented calcite crystal and has equal chances of being detected in the outgoing vertically polarized or horizontally polarized beam.

That is, nature has events that are truly random. It isn't simply a matter of not knowing enough about the situation. By knowing the photon is D-pol before it enters the crystal, you know everything there is to know about that photon's polarization, according to quantum theory. Yet the outcome of the measurement is unpredictable and therefore considered random. This fact bothered some scientists and philosophers who were trained in classical physics, in which everything is assumed to be predetermined and predictable, at least in principle. Adopting that view of nature, it was natural to think randomness appears only because you lacked some information about the situation, so you couldn't make perfect predictions.

We now know this is not true. Nature has certain events that are fundamentally and intrinsically unpredictable— for example, a single D-pol photon being measured with an H/V crystal. Quantum theory does allow us to calculate precise probabilities for these events, and these probabilities give us great predictive power when many such events are considered together. Consider a beam of light consisting of many photons that all have D polarization. Let's say we use an H/V-oriented calcite crystal to analyze this beam of light. Then, roughly half the photons are registered at the V-pol detector and half at the H-pol detector, giving equal light power at each detector. This is the same result we predicted earlier using the classical theory of light. The classical and quantum theories are consistent in their predictions in this case.

Not all quantum events are unpredictable. For example, if a single D-pol photon is measured using a D/A calcite crystal (as in FIGURE 2.5), we can predict the outcome with one-hundred-percent certainty: The photon will arrive at the D-pol detector. The same is true for an H-pol photon passing through an H/V calcite. In fact, for any purely polarized photon with known polarization, there is always one particular measurement you could make for which you can predict the outcome perfectly: simply line up the polarizer orientation with that of the photon.

How can we prepare a photon with a particular pure polarization?

Let's say you wish to have a photon with pure V polarization. To prepare such a photon, send a photon with unknown polarization into an H/V-oriented calcite crystal, as in FIGURE 2.3. Remove the V-pol detector from the setup and observe the H-pol detector. If you know the photon has passed through the crystal but has not been registered by the H-pol detector, then you know it is now traveling in the upper beam where there is no detector, and has pure V polarization. On the other hand, if you do observe a detection event at the H-pol detector, then the method has failed. So, try again with another photon, and keep trying until you succeed.

Can you determine the polarization of a single photon by quantum measurement?

No, you cannot, and this is an essential point. Imagine a photon for which you have no information about its polarization.

Say you measure its polarization using a calcite crystal that is oriented H/V. Say you obtain H-pol as the result. All you have learned is that the photon wasn't exactly V-polarized; it could have had any other polarization.

In fact, quantum theory and experiments show there is *no* device that can determine the polarization of an individual photon. The nonexistence of such a device provides the basis for quantum cryptography—a method for sending messages across the Internet with near-perfect security. This is the subject of Chapter 3.

What is the difference between the classical and quantum concepts of polarization of light?

At the beginning of this chapter, I mentioned three aspects of the measurement process that common sense and classical physics entice us to believe:

1. We should be able to increase the precision of measurements without limit.
2. We should be able to make measurements in a way that does not significantly change the objects being measured.
3. The quantity we wish to measure actually *has* a value before we perform a measurement, and the act of measuring merely reveals this value.

From the previous examples with single photons, we begin to see there are major differences between the measurement of quantum objects and the concept of classical measurement. First, there is no measurement we can perform to determine the previously prepared polarization of a single photon if we don't know it in advance. This illustrates a limitation that exists in the precision of a measurement of a quantum object. Second, when we attempt to measure the photon's polarization, the measurement process can actually change the photon's polarization, as illustrated in FIGURE 2.8. This change seems to be at random, with no underlying mechanism that we know of. We can categorize a beam of photons by the probability that each photon in the beam will follow the V-pol or the H-pol beam path, which in the case considered earlier is 50/50. But in general, we cannot predict with certainty what any single photon will do! We must deal with probability when dealing with quantum objects.

This unpredictability is a fundamental quantum feature of nature, which is unavoidable in the quantum realm. It is a great departure from the behavior of objects in the human scale or classical realm, where it is assumed that every event is perfectly predictable if only we have enough detailed information about the object.

How do we predict probabilities for photon polarization measurements?

Probability is a number representing your confidence that a given event will happen (or has happened in the past). For example, if you are forty percent confident it will rain tomorrow, you say, "The probability it will rain equals 0.4." Here I give a method for predicting probabilities for the outcomes of photon polarization measurements using only simple geometry and concepts from quantum theory.

Recall that a D-pol photon, if measured using an H/V-oriented calcite crystal, has a 0.5 probability to be registered at the detectors as H, and the same for V. But what would you expect for these H and V probabilities in the case when the photon's polarization is oriented just a few degrees from vertical? It might seem reasonable to expect that the probability for detecting V is much greater than for detecting H. This is correct, and the intuitive reason is that this photon's polarization is more 'similar to' the V direction than it is to the H direction.

We want to turn this idea of 'similar to' into a probability, which is a number. According to quantum theory, you can do this just by making drawings and analyzing the geometry. FIGURE 2.9 shows such a drawing for the case of a D-pol photon. On the left is a rear-end view of the light beam traveling away from you, and the tilted line represents the polarization. The photon's polarization direction is labeled D-pol. The calcite crystal used to analyze this photon has a unique axis, which is oriented in the V direction. In this case, the polarization is equally similar to the H direction and the V direction, so we expect the probabilities of these outcomes to be equal. We also know the two probabilities must add up to 1, so each probability must equal 0.5.

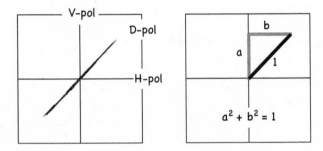

Figure 2.9 A single photon's diagonal polarization, and its representation as the long side of a right triangle.

In the figure on the right, there is a triangle with the two short sides in the V and H directions, and the long side in the direction of the photon's polarization. The length of the polarization equals 1, to reflect the fact that the probabilities must add to 1. The short lines are the *components* of polarization along the directions V and H. Their lengths are indicated using the letters a and b. Notice that the three lines form a right triangle. We know from the Pythagorean Theorem that for a right triangle, the sum of the squares of the two short sides equals the square of the long side. Because $1^2 = 1$, we have the relation $a^2 + b^2 = 1$.

Here we make a conjecture, which was first made by physicist Max Born in 1925. The idea is that we want to arrive at a formula containing two probabilities that add to 1. For this reason, Max Born identified a^2 as the probability for the photon to exit in the V-pol beam, and b^2 as the probability for the photon to exit in the H-pol beam. This identification is called *Born's Rule*. It has been found to be in perfect agreement with every experiment ever done concerning the polarization properties of photons.

Born's Rule is easy to apply simply by drawing figures and using a ruler: Redraw the triangle shown in the figure on a sheet of paper such that the length of the polarization line is 1 meter. Let's refer to this length as 1, without including the word 'meter.' Then, if you use your ruler to measure the lengths of the other two sides, you will find lengths a and b both close to 0.707. If you square this number, you will find $0.707^2 = 0.5$. So, according to Born's Rule, the probabilities to observe detection events in the V and H beams both equal 0.5. This is in accordance with our earlier expectation that the odds are 50/50 the photon will take either path. So Born's method works in this case.

Consider another example. Let's say a photon is prepared having a polarization known to a protagonist named Alice. Let's say that the polarization direction makes a 30-degree angle with the H direction, as in FIGURE 2.10. Alice verifies the polarization angle by passing the photon through a calcite crystal oriented at 30 degrees from H. The photon passes the test. Now she sends that photon to another protagonist named Bob, who does not know the photon's polarization angle. (We encounter Alice and Bob many times in this book, so let's get acquainted with them.)

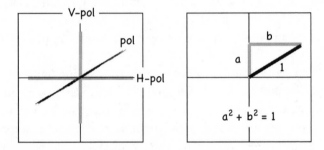

Figure 2.10 A photon has polarization direction (pol) oriented at 30 degrees from the horizontal direction. Probabilities add to 1.

To measure this photon, Bob chooses to use a calcite crystal in the H/V orientation. How do we predict the probability that the photon will exit this crystal in the H-pol beam? Perhaps you can determine intuitively which measurement outcome will have the higher probability. Which polarization, V or H, is the photon's polarization most similar to? If you thought H, you are correct; the polarization is tilted more toward H than toward V. How can we use Born's Rule to determine the value of the probability?

Shown in FIGURE 2.10 is the right triangle made up of lengths a, b, and 1. If you use a ruler to measure length b carefully, you find a value close to 0.866 in this example. The probability for the H-pol outcome is then $0.866^2 = 0.75$—that is, a seventy-five percent chance the photon takes the H-pol beam path. If you use a ruler to measure length a carefully, you find a value close to 0.50. The probability for the V-pol outcome (that is, perpendicular to the H-pol) would therefore be $0.50^2 = 0.25$—that is, a twenty-five percent chance the photon takes the V-pol beam path. All is well, since $0.75 + 0.25 = 1.0$, as it must be.

We will see in later chapters that the Born Rule applies to many situations in the quantum realm. If you followed the reasoning in this section, you are now an expert in an important part of quantum theory! If not, no worries! Let's move onward.

What does it mean to make a measurement in the quantum realm?

To summarize the previous discussions, all experiments on measuring photon polarization involve three parts: (1) the preparation of a photon, (2) the choice of calcite crystal orientation, and (3) the occurrence of a particular *measurement outcome* at a detector.

A few important comments about these steps:

- There are infinitely many options for the choice of calcite crystal orientation to be used to analyze a photon's polarization, and for a given measurement you can choose only one of them. I will call the chosen crystal orientation the *measurement scheme.*
- For a given measurement scheme, there are only two possible outcomes, the probabilities of which add to 1.
- There always exists one choice of measurement scheme for which the outcome can be predicted with one-hundred-percent certainty; that is, when using this special scheme, one of the outcome probabilities equals 1, whereas the other equals 0.

So what does it mean to make a measurement in the quantum realm? Many deep-thinking physicists have tried answering this question. To quote one, Sir James Jeans in 1943 wrote[1]:

> Our minds can never step out of their prison-houses to investigate the real nature of things—gold, water, centimeters or wavelengths—which inhabit that mysterious world out beyond our sense organs. We are acquainted with such things only through the messages we receive from them through the windows of our senses, and these tell us nothing about the essential nature of their origins.

Jeans is getting at the fact that any measurements made on a single photon cannot determine its polarization, yet a particular measurement does yield some definite outcome. The measurement outcome does not correspond directly to a past, definite property of the photon. From such a measurement we can receive only a limited amount of information about the photon's polarization. Again, if we detect the photon in the V outcome detector of an H/V calcite crystal, the only thing we can say with certainty is that the photon's polarization was not H. As Jeans says, it is as if the photon is sending us a 'message' that contains some, but not complete, information about the question we asked.

What is measurement complementarity?

The outcomes that are possible from an observation of a photon's polarization depends on how you choose to set up your experimental apparatus—that is, your measurement scheme. For example, you can set up your calcite crystal to find out whether a photon yields the outcome V-pol or H-pol. Or you can set up your crystal to find out whether a photon yields the outcome A-pol or D-pol. But, for a given photon, you cannot do both! You can't turn back the clock and proceed as if you hadn't made the first measurement.

You might think: Well, I can make a sequence of measurements; first measure using the H/V scheme, then measure using the D/A scheme. The trouble with this proposal is that the first measurement actually changes the polarization of the photon in an uncontrollable way, which defeats your goal. FIGURE 2.11 shows such a setup. There is a detector labeled 'V-detector' in the V-pol path; if that detector does *not* register an event, then you know the photon took the H-pol path. When it reaches the D/A-oriented crystal, according to Born's Rule, it then has a 50/50 chance to be registered at either the A-pol or D-pol detectors. But this 50/50 chance has nothing to do with the original polarization of the photon, because that was changed during the H/V measurement!

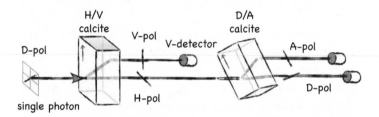

Figure 2.11 A sequence of measurements does not yield two independent pieces of useful information about a photon's polarization.

Niels Bohr, a highly influential Danish physicist in the 1920s, introduced the concept of *measurement complementarity* to describe the fact that, for a given photon, you can't do both measurements H/V and D/A independently. He said that the V-or-H-pol observation is 'complementary to' the D-or-A-pol observation. Bohr's idea of measurement complementarity applies not only to photon polarization orientations, but also applies, for example, to the position and velocity of an electron. They cannot both be observed independently for a single electron.

How can a human-scale object seem to possess definite
properties if the individual quantum objects making it up do not?

We can't determine the polarization orientation of an individual photon, but if the beam is comprised of a very large
number of photons, each with a possibly different polarization
orientation, we can determine the average polarization orientation of the beam. We can do this by rotating slowly a calcite
crystal that is placed in the beam and, for each angle, recording
the fraction of photons that exit in either beam from the crystal.
Let's say we find a crystal orientation for which the maximum
fraction of photons is detected in the lower beam exiting the
crystal. This particular crystal orientation is the average polarization of the beam. If we consider the light beam as a whole to
be a human-scale object, then we can say we have determined
its polarization.

Although individual quantum objects don't possess clearly
describable properties, when a large collection of quantum
objects come together to create something that we humans
perceive as a human-scale object, then our human intuition
allows us to assign definite values to properties we ascribe
to that object. We perceive that a stone thrown in the air has
definite values of position and velocity at any instant of time.
The fact that a stone follows a definite trajectory is common
sense, and there is no reason to doubt it. Yet the stone is made
up of a very large number of quantum objects (electrons, protons, neutrons) for which the measurements of position and
velocity would, individually, yield fundamentally random
outcomes, just as the polarization does for an individual photon. So what do we really mean by the concept of the position
of the stone? We mean the *average* position of the large collection of objects making up the stone. This average value has
very little randomness associated with it, because the randomness of its individual constituents averages out.

This idea might help you understand what I meant earlier
when I referred to classical and quantum 'realms.' This, of

course, does not mean there are two different kinds of universes that coexist. There is only one world, and we live in it. By classical realm I mean those aspects of the world you can touch or see with your own human perceptions, and the properties of which can be measured without changing them significantly. We then perceive that they obey the laws of classical physics. By quantum realm I mean those aspects that can be perceived only indirectly because they correspond to elementary quantum objects. Measuring them changes them. The classical realm appears to emerge from the quantum realm when we focus our attention on a particular human-scale object or quantity that can be touched or seen using human senses.

What do we mean by the state of an object?

The word 'state' connotes condition, such as when we refer to a person's state of mind. In physics, if we know the state of an object, then we know everything possible to know about that object. What is a state in classical physics, which itself is an idealized description of nature that applies to human-scale objects? In classical physics, we can know and specify precisely the position and velocity of an object, such as a stone. There might be other aspects or properties to specify, such as rotation rate (is a baseball spinning?). The full specification of all of these aspects is called the *classical state* of an object.

In classical physics, if we don't know the state of an object we can determine it by making careful measurements of its position, velocity, rotation rate, and so forth. In classical physics, the state of the object determines precisely and uniquely the outcome of any measurement you could make on the object. So in classical physics, the state corresponds directly to the collection of all possible measurement outcomes, all of which can be predicted with certainty if the state is known.

What is a quantum state?

What is the meaning of 'state' when referring to quantum objects? Again, if we know the state, then we know everything possible to know about that object. But there are limits to how much we can know about quantum objects. We cannot know with certainty what will be the outcome of every measurement we can make on the object because, in most situations, these outcomes are fundamentally random. Quantum theory, however, does allow us to predict the probabilities for each possible outcome for any measurement you might choose to make on the object. So we say the *quantum state* corresponds to the collection of probabilities for outcomes for all possible measurements you can make on the object. In contrast to classical physics, wherein the state corresponds directly to the actual measurement outcomes (which are certain if you know the state), in quantum physics the state corresponds to the probabilities for outcomes (which typically are not certain even if you know the state).

In our discussion of photons, I pointed out that you can prepare the polarization orientation of a single photon, and so it can be known before any measurements are made. This known polarization orientation of the single photon is an example of a quantum state. It does not represent any preexisting inherent property of the photon, such as position and velocity represent for a stone. Nevertheless, it represents everything that is possible to know about this particular photon's polarization, and this knowledge is independent of the experimental scheme we might adopt for making measurements.

In our earlier example, Alice prepared the photon, so she knew its quantum state. But, when she passed the photon to Bob, there was no way he could make measurements on this single photon to determine its quantum state. The impossibility of determining the state of a single quantum object illustrates another big difference between classical physics and quantum physics. It may seem to be a bother conceptually, but it is a boon practically, because we can use this feature to build perfectly secure communication systems, as discussed in Chapter 3.

Earlier we represented the state of a single photon by drawing a line in a certain direction, as seen in FIGURE 2.10. We saw that by using Born's Rule, we can predict the probabilities of any outcome for a particular experimental measurement scheme. Here, then, is a useful definition of a quantum state:

A *quantum state* is a representation of everything possible to know about a particular quantum object, permitting one to calculate probabilities for fundamentally random observation outcomes in any chosen measurement scheme.

In other words, a quantum state refers to as complete a description as can be given about a quantum entity or object.[2] If we know fully the quantum state of an entity such as an electron or photon, there really is no more information to be had.

In the case of a single photon, we refer to the polarization orientation as the *quantum state of polarization*, whereas in the case of a strong light beam, which can be described using classical physics theory, we instead refer directly to the polarization of the light as a property of the light. This subtle difference in thinking is needed because of our inability to determine the polarization of a single photon. For a single photon, you can have only one try at a polarization measurement, giving only a small amount of information. In contrast, a strong light beam can be measured repeatedly, giving more and more information about its polarization state. Thus, the idea of measurement in the classical realm is an idealization that is valid for human-scale objects. For such objects it is intuitive to believe in the possibilities of unlimited measurement precision, measurement without disturbance, and the preexistence of measurement outcomes before any measurement is made. These concepts are behind the concept of classical state.

Can Bob determine a quantum state experimentally that was prepared by Alice?

Although Bob cannot determine the quantum state of an individual photon prepared by Alice, there is a case in which he can determine a quantum state by making observations. If Alice prepares many (say, 10,000) individual photons all by the same procedure, so she knows the quantum state is common to all photons, then Bob can design an observation strategy that reveals to him the state Alice prepared.

To make the example simple, let's say Bob knows (because Alice informed him truthfully) that the polarization orientation of all the photons is pointing along the same orientation (but is unknown to Bob).[3] To determine this orientation, which specifies the quantum state, Bob can divide his observations into two sets. For the first 5000 photons, he can rotate his calcite crystal into the V orientation and observe the fraction that passes into the V-pol outcome beam. This tells him the angle the photons' polarization orientation makes with the V direction. But it does not tell him if the angle is positive (clockwise) or negative (counterclockwise) relative to the V direction. To break the tie, he can use the second 5000 photons. He can simply guess that the correct orientation is positive (clockwise) relative to the V direction, then rotate his crystal into that orientation and observe what happens. If his guess is correct, he will see one-hundred percent of the photons being detected in the lower beam emerging from that crystal. If his guess is incorrect, he will observe the opposite behavior. In either case, he now knows the correct orientation of the photons' polarization state. This specifies completely the quantum state of the photons' polarization.

The take-home message is there is an observation strategy that reveals to Bob the state Alice prepared, but only if Alice prepares many photons identically and passes them to Bob. Measurement schemes such as this are called *quantumstate tomography*, because, as in medical imaging—such as computer-aided tomographic (CAT) scans—the object is hidden from direct view, but its nature can be revealed indirectly by making a series of different measurements. (In a CAT scan, X-ray images are collected at many angles and a set of images is used to reconstruct a view inside the body.)

Can Bob make copies (clones) of the state of a single photon?

In the classical realm of human-scale objects, there is no problem with making copies or replicas of any individual object: an artwork, a page of text, and so on. To do this, we need to examine every aspect or property of the original object. But, in the quantum realm, we cannot examine every aspect or property of an individual object because some of them are complementary to each other, as Niels Bohr would say. That is, if you measure a certain property, then there are other properties that you can't measure meaningfully. This defeats the most obvious way of making a perfect copy of a quantum object, including its quantum state.

Perhaps we could make a perfect copy of an object including its quantum state if we did *not* try to determine that state, but used some kind of automatic quantum-state cloning machine. Perhaps such a machine could work even without determining the state itself. Alas, one can prove from quantum theory that such state cloning, or copying, is impossible without first destroying the original object. You can't make two from one. The quantum copier machine is permanently out of order! This is an important point, and it plays a role in many applications, such as data encryption, as discussed in Chapter 3.

This *no-cloning principle* also reinforces the conclusion of the previous section in that the quantum state of a single object cannot be determined by any measurements. To see this, think what would be possible if we could clone the state of a single object. We could make many copies of that object's state, then follow the quantum-state tomography procedure described earlier for determining a quantum state, given the availability of many identically prepared objects. This would allow us to determine the state of the single original object, in disagreement with what is allowed by quantum physics.

What is quantum coherence?

Previously we discussed coherence in the context of classical physics. Coherence in the context of quantum physics, or *quantum coherence*, is analogous to coherence in classical polarization, except instead of referring to an aspect of the electric field, as in the classical case, quantum coherence refers to an aspect of the quantum state. That is, the two components, a and b, of the quantum state must be stable and not wander. Only in this case does Born's Rule yield the correct predication for probabilities. Physicists use the term *coherent superposition* to describe a combination of two state components that give rise to a particular quantum state.

What are the Guiding Principles of quantum mechanics?

Based on the material covered in the previous sections, physicists have drawn some general conclusions about how quantum objects behave when a person attempts to observe or measure them. I present these conclusions as Guiding Principles of Quantum Mechanics. These principles largely replace the three aspects of the classical concept of measurement with which this chapter began. Here are the first three quantum principles, based on our earlier discussions (others are stated in later chapters):

Guiding Principle #1 The world is intrinsically probabilistic. Even if you have maximum knowledge about an object, you cannot predict with certainty the outcome of a typical measurement of that object.

Comment: Unlike our conception of classical physics, quantum physics makes it clear that the idea of probability is fundamental to understanding and describing nature. This does not mean we need to give up our intuitive ideas of cause and effect. Max Born wrote, "[In classical physics] one physical situation depends on the other. This is still true in quantum physics, though the objects of observation for which a dependence is claimed are different: they are the probabilities of elementary events, not those single events themselves." He is saying that one can predict and know probabilities, but not predict with certainty specific outcomes.

> **Guiding Principle #2** The condition of a quantum object can be described by specifying its quantum state—a mathematical representation of everything that is possible to know about a particular quantum object, permitting one to calculate probabilities for observation outcomes in any chosen experimental situation.

Comment: The outcomes of observations should not be thought of as measurements of preexisting properties of the object. Rather, those outcomes provide partial information about the underlying state of the object. If many identically prepared objects are available for study, then by using a clever strategy you can obtain complete information about the quantum state common to all the objects. But this is not possible if only one such object is available.

Comment: If you have access to many objects that are *not* prepared identically, then the whole collection cannot be characterized by any perfectly known quantum state. In such a case, it suffices to characterize the 'mixture of states' in which the collection of objects is prepared. Such a mixture of states can be characterized experimentally by carrying out a series of measurements using different measurement schemes.

> **Guiding Principle #3** Certain pairs of observed attributes are physically incompatible or complementary to one another, in that they cannot both be observed independently. More generally, the results of measurement depend on the particular measurement scheme with which it is performed.

Comment: You cannot answer the question "Is a single photon V-pol or H-pol?" and independently answer the question "Is it D-pol or A-pol?" These are incompatible questions. One can say the purpose of measurement is not to determine the values of any preexisting properties, but simply to find out what the outcome will be if you choose to observe a particular aspect in a particular experimental setup. For this reason, observed results are simply called *measurement outcomes*. This wording does away with any implication that the outcome was predetermined by any preexisting properties that the object possessed before the measurement.

Let's also emphasize that any experimental apparatus used to determine an observed aspect in a particular experimental context will typically change the quantum state of the object. For example, if a D-pol photon enters a calcite crystal and exits in the H-pol beam, then its polarization state is now H-pol. This, then, fixes the probabilities for any further observations of polarization.

What does quantum mechanics really describe?

I will answer this question with a quote from Max Born: "Thinking in terms of quantum theory needs some effort and considerable practice. The clue is . . . that quantum mechanics does not describe a situation in an objective external world, but a definite experimental arrangement for observing a section of the external world." He goes on to say, "our sense impressions are . . . indications, or signals from, an external world which exists independently of us."[4]

Notes

1　Sir James Jeans, *Physics and Philosophy* (Cambridge: Cambridge University Press, 1943; reprinted by Dover, 1981), 8.

2　This paraphrases a discussion of quantum states by Lee Smolin in *Three Roads to Quantum Gravity* (New York, NY: Basic Books, 2002)—a recommended book for the general reader.

3　For simplicity, we are not considering the possibility of so-called circular polarizations.

4　Max Born, *Natural Philosophy of Cause and Chance* (New York, NY: Dover, 1964), 103 and 108.

3

APPLICATION: QUANTUM DATA ENCRYPTION

Can quantum physics be harnessed to create perfectly secure Internet communication?

Yes. In recent decades scientists have learned how to use the quantum nature of light to make messages and other data almost perfectly secure. Quantum physics allows us to encrypt messages in ways that cannot be cracked by any computing method in a reasonable amount of time. Data encryption using quantum techniques is an excellent topic for our first foray into technological applications of quantum physics.

How does encryption keep messages secret?

For financial, industrial, military, or personal reasons, people often want to send messages so that only the sender and intended recipient can know its contents. Typically, encryption systems use some kind of 'replacement system,' in which letters or numbers are replaced by other letters, numbers, or symbols according to some preset, but secret, rule. The more complex the rule, the harder the system is to crack.

The art of sending messages in secret has a history going back many centuries. In ancient times, Julius Caesar used a system, now called the Caesar Cipher, in which each letter of the alphabet was replaced by the letter appearing some fixed number of positions later in the alphabet (and starting over at the beginning when the alphabet's end is reached). The fixed number is called the *encryption key*, because it is used to 'lock' and 'unlock' the encrypted text. For example, if the key equals 2, the text, "I love you" would be replaced by "K nqxg aqw." The recipient, if he knows the key equals 2, can easily reverse, or decipher, the encryption process by shifting each letter back by two positions in the alphabet. If the key is known only to the sender and the intended recipient, it is called a secret key. A third person, lacking the key, will have a tough time deciphering the message.

Can most encryption methods typically be cracked?

Yes. With enough ingenuity, an adversary can always find a way to decipher or 'crack' your secret code, no matter how complex or tricky the rule, if it relies on a fixed replacement rule used repeatedly. For example, let's say the sender, Alice, sends the encrypted message "K nqxg aqw" to Bob, the recipient, who knows the key equals 2. He can decipher the message easily and read it. (I hope he's pleased! But here we are not concerned with the meaning of the messages, only whether their information content can be determined.) Now suppose an eavesdropper, Eve, intercepts this message, copies it, and sends it on its way to Bob, hoping he and Alice are none the wiser. Can Eve decipher it? In this case, probably yes. She simply has to guess that a Caesar Cipher is being used, and then try each of 26 possible keys and determine whether any yields a sensible-looking message, rather than gibberish, which she thinks is most likely the correct one. Notice she cannot be completely sure her guess is correct. Even when she sees the deciphered text "I love you," she must decide whether this is most likely the correct message. The longer the message, the less likely a deciphered message would accidently contain a sentence of intelligible English. For this message of eight letters, she would feel fairly confident that "I love you" is the message intended by Alice for Bob.

To try to defeat eavesdroppers, more and more sophisticated versions of such replacement codes have been used since

Roman times, and they were always cracked eventually by using more and more powerful mathematical techniques. That is, when using letter replacement methods, one never knows if a clever adversary has worked out a method to crack your secret encryption method. A recurring situation in history was that senders and receivers developed secret methods they believed to be so complex, subtle, or tricky that no eavesdropper could possibly guess them or figure them out—only to be surprised later to learn their adversary was able to do just that.

A famous example (dramatized in the movie *The Imitation Game*) occurred during World War II, when the Nazis invented and used a mechanical message-encrypting device called the Enigma machine, which they believed could create secret codes that could not be cracked by the Allies. In response, Polish mathematicians and, later, British researchers led by Alan Turing, worked hard to invent their own machines that could perform mathematical calculations quickly in ways designed to crack the Enigma codes. They deciphered military secret messages of the Nazis successfully, thereby influencing the outcome of the war. The machines and methods the Allies invented for this effort foreshadowed the computers of today, so there was some positive outcome in all of this otherwise dreadful history.

Is there an encryption method that cannot be cracked?

Yes. Think back to the message, "I love you." What if Alice were to shift each letter of the message by a different number of alphabet positions, and these shifts were known only by Bob? For example, shift the first by 2, the second by 12, the third by 9, and so on. Bob can decipher this message. But now Eve has the problem that the original text could be absolutely anything! There is absolutely no way for Eve to crack such a method.

Unfortunately for Alice and Bob, though, the list of key numbers (2, 12, 9, ...) needs to be just as long as the message itself. Furthermore, they need to be clever enough never to use the same key list twice. That is, they must use a given key list, now called simply the 'key,' only once and then throw it away. Imagine that Alice, in advance, has given to Bob a key consisting of a list of 10,000 key numbers written on the pages of a notebook (or 'pad'). He and Alice could use this pad of numbers only once. For this reason, the large list of key numbers is called a 'one-time pad.' For example, if you want to send a ten-page document of text, you need to use a separate ten-page-long key to encrypt and decrypt it. This is one purpose of the diplomatic pouches that are hand-carried between a country's embassies. They may contain computer hard drives (or, in the old days, printed books) containing random keys, which can be used later to send and receive perfectly secure encrypted messages.

What if you run out of shared keys and are unable to share new ones through a method such as a diplomatic pouch? Can you send a new set of keys to your legitimate partner without having the key data itself intercepted by your adversary? Using standard methods, the answer is no. This is where the unique features of quantum physics come to the rescue. Before discussing quantum methods for key sharing, we need to understand how text is represented using binary encoding.

How is text represented using binary symbols?

Computers store and manipulate information using a special language with an alphabet that consists of only two symbols: 0 and 1. This language is called *binary*, and each symbol is called a *bit*, short for binary digit. Because all modern data transfer and encryption systems use computers, we need a method to translate our text files into binary. This is done with a simple lookup table, which is used to convert each lowercase or uppercase letter in the Latin alphabet into a unique sequence of eight bits. For example 'A' becomes 01000001, 'B' becomes 01000010, and so on, and 'a' becomes 01100001, 'b' becomes 01100010, and so forth. You can easily find the entire lookup table for this system, which is called ASCII, by searching the Internet. There is nothing secret about this method.

How is a text message encrypted and decrypted using a binary key?

To encrypt a text message represented as a long sequence of 0s and 1s using the ASCII system, a legitimate party named Alice can use an equally long sequence of randomly generated 0s and 1s that serve as the binary encryption key. The rule is as follows: For a given symbol position in the message, if the key symbol is 0, leave the corresponding text symbol as it is; if the key symbol is 1, flip the corresponding text symbol from 0 to 1 or from 1 to 0. This creates the encrypted message. For example, if the original message is 11110000 and the key is 10101010, then the encrypted sequence is 01011010. Alice can then transmit the encrypted sequence over a nonsecure channel to the intended recipient, Bob. If Bob has the secret key, he can reverse the encryption procedure easily and recover the message. The overall system is illustrated in FIGURE 3.1.

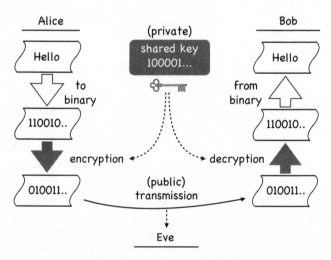

Figure 3.1 Alice sends an encrypted message to Bob using a secret shared key.

This seemingly simple method is completely secure, as long as Alice and Bob possess the same secret key: a sequence of randomly generated 0s and 1s, or bits, of length equal to that of the message. If they don't share such a key initially, the challenge is for Alice to be able to send to Bob a long binary key before they want to use it to encrypt a message.

How can photon polarization be used for creating secure encryption keys?

For Alice to share a secret key with Bob, she can use single photons to send bits (0 or 1) to Bob, with each photon carrying one bit. Recall that in a quantum physics description, light consists of discrete units of energy called photons, the polarization of which can be manipulated and measured. The security of this method rests on the fact that photons are elementary quantum entities, so measurement cannot be considered as simply passive revealing of an object's state but is a fundamental quantum process. As such, if an eavesdropper intercepts a photon and measures the bit value it carries, the photon's polarization is necessarily disturbed. Bob and Alice can detect this disturbance, thus revealing the presence of an eavesdropper.

The goal of such systems is to allow a user to define and send, or 'distribute,' a secret key to an intended recipient. This step is called *key distribution,* and it must be secure, although the communication channel used is insecure in the sense that eavesdroppers may have access to it. The method of encryption used by Alice and Bob can be known to everyone; it need not be a secret. The security of the system relies not on its complexity, but on the physical nature of quantum systems.

What physics principles underlie quantum key distribution?

To understand the physics underlying the procedure of sending or distributing a key secretly, you need only recall a few facts about measuring photon polarization from Chapter 2. First, light behaves in some respects as a wave in which the electric field oscillates as the wave travels. Light has a property called polarization; the electric field can point in any direction perpendicular to the direction of the wave's travel. For our purposes, we consider only four relevant polarization directions: horizontal (H), vertical (V), diagonal (D), and antidiagonal (A), and they are abbreviated as H-pol, V-pol, D-pol, and A-pol.

Second, energy in light is carried and detected as discrete units, called photons. To measure a photon's polarization, we can use a calcite crystal followed by two light detectors. The crystal splits the light into two beams, depending on polarization relative to the crystal's orientation, and only one of the detectors registers a detection event for each incident photon. If the crystal is oriented to split H- and V-polarized light, we say we are using the H/V measurement scheme. If the crystal is oriented to split D- and A-polarized light, we say we are using the D/A measurement scheme.

Third, if a photon is prepared in the H-pol state and is measured in the H/V scheme, it will certainly be detected in the H beam path, not the V path. If the same photon is measured in the D/A scheme, it is equally likely (with fifty-percent probabilities) to be detected in either of the H or V beam paths.

Fourth, the polarization quantum state of an individual photon cannot be determined by any measurement scheme. Consequently, perfect copies of an individual photon cannot be made.

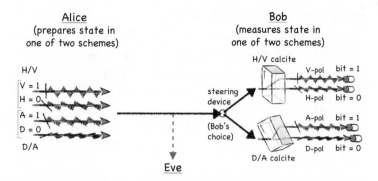

Figure 3.2 Alice may send a given photon prepared in one of four possible polarizations; V-pol or A-pol represents bit value 1, and H-pol or D-pol represents bit value 0. Bob randomly chooses to steer each photon to one of two polarization analyzers, each of which yields a measurement in one or the other polarization scheme.

As illustrated in FIGURE 3.2, Alice prepares photons with specific polarizations and sends them to Bob. She uses four possible polarization states of photons to represent the key data. While preparing a sequence of photons, Alice switches randomly between the two state schemes in order to confuse an eavesdropper, Eve, who might be monitoring those photons. If Alice chooses the H/V scheme for a given photon, then H represents a key number 0 and V represents a key number 1; if, instead, she chooses the D/A scheme, then D represents a key number 0 and A represents a key number 1. For example, if she transmits photons in the sequence HHDVADAV, it represents a segment of a key given by 00011011.

If Bob could measure each received photon using the same scheme Alice used for that photon, he would receive the correct bit value. The challenge is to do this in such a way that an eavesdropper cannot know which key data were transmitted. This is possible because an eavesdropper does not know (nor does Bob) which polarization scheme Alice used to send each photon. The key is constructed from the secret lists of 0s and 1s that Alice and Bob have, after all the photons have been received by Bob, as explained next.

How does quantum key distribution work?

Before Alice can send Bob a secure message using a binary key as illustrated in FIGURE 3.1, they must share the key. Recall that the bit values 0 or 1 are the actual polarization states of the sent photons, whereas the 'schemes' refer to whether these states are distinguished as H/V or as D/A. The procedure for creating such a private key is illustrated in FIGURE 3.3, and has the following steps:

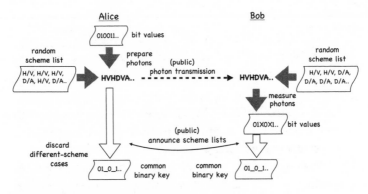

Figure 3.3 Procedure for creating a private key.

1. Alice creates a long random list of bit values—0s and 1s—some of which will end up being part of the shared key number. Say she generates

   010011101100001101011111100101. . . .

2. Alice uses some random procedure (such as coin flipping) to decide in which scheme she will prepare each photon, and then she records this scheme list. Say she generates the scheme list H/V, H/V, H/V, D/A, H/V, D/A, D/A, D/A, H/V, D/A, and so on.

3. Alice creates and prepares photons sequentially each in the state corresponding to the bit value using the proper scheme—H/V or D/A—according to her scheme list. Recall H and D represent 0, whereas V and A represent

1. Only Alice knows these states and schemes. (We have to presume her workstation is physically secure; no spies are present.) Alice transmits these photons to Bob to generate a secret shared key number.

4. Bob, who doesn't know Alice's bit values or scheme list, measures each incoming photon, randomly picking a measurement scheme to use for each photon. He records both the list of the schemes he used and the photon polarizations—H, V, D, or A—observed for each photon detected.

5. For the moment, let's consider the case that no eavesdropper is present, so that all photons are transmitted faithfully. Then, Alice and Bob can generate a shared key number by the following method: They announce publicly their preparation scheme and measurement scheme lists, so they and anyone else can now know what schemes they used earlier for sending and receiving each photon already transmitted. But Alice keeps her sent bit values secret, and Bob keeps his measured bit values secret.

For about one-half of the photons transmitted, Alice and Bob have, by chance, used different schemes—say, H/V for Alice and D/A for Bob, or vice versa. They must discard these cases, because there is no correspondence between the photon state that Alice sent and the photon polarization that Bob measured. This is a consequence of the quantum physics principles discussed earlier.

In contrast, for the other half of the cases, in which Alice and Bob used the same scheme, the bit values will match perfectly (when there is no eavesdropper), according to the workings of quantum mechanics. So, if they keep only those cases, they will now possess identical bit lists, which can be used as a shared secret key. These lists can be used to encrypt and decrypt messages they wish to send to each other (until they run out of unused key bits).

An example of the procedure is shown in Table 3.1 (presuming there is no eavesdropper). Alice's attempts are labeled a through j. In the fifty percent of cases for which Alice and Bob used the same scheme (which they learn only after the fact), their bit values always agree. In cases for which Alice and Bob used different schemes, Bob's measured bit value can yield 0 or 1 with equal probability. After discarding those cases in which different schemes were used, they are left with the shared key bits listed in the far right of Table 3.1, which are in perfect agreement between Bob and Alice. To create a shared key containing, say, 5000 bits, Alice needs to send about 10,000 individual photons.

Table 3.1 Example series of photon transmissions for quantum key distribution with no eavesdropper. H and D represent 0; V and A represent 1.

		Alice			Bob		
	Bit	Scheme	Transmit		Scheme	Bit	Key
a	1	H/V	V	→	D/A	0,1	
b	1	D/A	A	→	D/A	1	1
c	0	D/A	D	→	D/A	0	0
d	1	D/A	A	→	H/V	0,1	
e	1	H/V	V	→	H/V	1	1
f	0	D/A	D	→	D/A	0	0
g	1	H/V	V	→	H/V	1	1
h	0	H/V	H	→	D/A	0,1	
i	0	D/A	D	→	H/V	0,1	
j	0	D/A	D	→	D/A	0	0

What if an eavesdropper is present?

Say Eve the eavesdropper is present somewhere between Alice and Bob. Table 3.2 shows that Eve intercepts each photon and guesses randomly which scheme Alice used to prepare it. Eve measures the photon and gets a bit value, which may or may not be correct, depending on her choice of scheme. Eve then creates a new photon with that same polarization and sends it on to Bob, hoping Alice and Bob don't notice this photon substitution. There are various cases that can occur:

- In about one-half of cases (a, c, d, f, g, i, ...), Eve guesses correctly the preparation scheme used by Alice (although Eve will not know she has guessed correctly). In these cases, she obtains the correct bit value, 0 or 1, when she measures the photon, so she can prepare and send a new photon to Bob with the same polarization as the photon Alice sent. Then, of these cases, half the time (c, f, g, ...) Bob also guesses the preparation scheme correctly and obtains the correct bit value. In these cases, Bob and Alice share the same bit value, but so does Eve.

- In the other half of all cases, Eve guesses incorrectly the preparation scheme used by Alice and sends a new photon to Bob (which is indicated with a slash through the transmission arrow in Table 3.2). And of these (a quarter of total cases), Bob sometimes guesses correctly the preparation scheme used by Alice (b, e, j ...). In some of these cases (b, j, ...), Bob obtains an incorrect bit value (boxed in Table 3.2). And in the other cases (e ...), purely by chance, Bob obtains the correct bit value (circled in Table 3.2).

Table 3.2 Example series of photon transmissions for quantum key distribution with an eavesdropper present.

	Alice Bit	Alice Scheme		Eve Scheme	Eve Bit	Send		Bob Scheme	Bob Bit	Key
a	1	H/V	V →	H/V	1	1 V →		D/A	0,1	
b	1	D/A	A →	H/V	0,1	0 H ⇸		D/A	0,1	[0]
c	0	D/A	D →	D/A	0	0 D →		D/A	0	0
d	1	D/A	A →	D/A	1	1 A →		H/V	0,1	
e	1	H/V	V →	D/A	0,1	1 A ⇸		H/V	0,1	(1)
f	0	D/A	D →	D/A	0	0 D →		D/A	0	0
g	1	H/V	V →	H/V	1	1 V →		H/V	1	1
h	0	H/V	H →	D/A	0,1	0 D ⇸		D/A	0	
i	0	D/A	D →	D/A	0	0 D →		H/V	0,1	
j	0	D/A	D →	H/V	0,1	1 H ⇸		D/A	0,1	[1]

- In the remaining quarter of total cases (h ...), both Eve and Bob guess incorrectly the preparation scheme used by Alice. Bob and Alice later discard these cases, so it doesn't matter what Eve's and Bob's recorded bit values are.

After all photons have been detected, Alice and Bob announce their preparation scheme and measurement scheme lists publicly, so they and anyone else can now know which schemes they used earlier for sending and receiving each photon already transmitted. Alice and Bob keep their bit values secret. For about half the cases, Alice and Bob find they used the same scheme, and so the bit values would match if there weren't an eavesdropper. But, when Eve is present, in cases like 'b' and 'j,' Bob end ups with a wrong bit value. So now there are two problems Alice and Bob must overcome: (1) some of their shared bits are in disagreement and (2) some of their shared bits are known by Eve. What can they do?

How can Alice and Bob detect Eve's presence?

Up to now, Bob and Alice have no idea if Eve is tampering, so they are at risk of using keys that don't agree and, worse, that Eve has partial knowledge of them. This is unacceptable. To check for the presence of Eve, Bob picks randomly some (say, ten percent) of the cases in which he and Alice used the same scheme and, using a nonsecure communication system, such as e-mail, informs Alice of the actual bit values he obtained for those cases. If they find all of their compared bit values agree, Alice and Bob can be confident no eavesdropper is present, and they can use the remainder (ninety percent) of the bit values as their perfectly secure shared key.

But, if they find that many of their publicly compared bit values do *not* agree, they will be sufficiently suspicious that Eve is present to motivate them to discard all the key bits. Perhaps they will try again and hope that Eve is taking a break from her duties, at which time they can try to create a shared key.

We have seen that when using *quantum key distribution,* the very act of an eavesdropper measuring a photon disturbs the photon's state in a fundamentally quantum mechanical way. In fact, in the simple method just explained, to measure a photon's polarization, Eve must destroy the photon by sending it into a light detector. When she then creates a new photon to substitute for the one she destroyed, it is often of the 'wrong' polarization compared with the original one Alice sent. This change allows Alice and Bob to detect Eve's presence.

This concludes the description of the basic principles and methods of key distribution using quantum physics. The following sections elaborate the method further. For example:

What if Eve is always present?

If Eve doesn't want to simply deny Alice and Bob the ability to share a key, but instead wants to fool them into thinking she is not eavesdropping, she might intercept only a small fraction of the photons, gaining a little information about the key that Alice and Bob are creating. Later, if Alice and Bob use the resulting key to exchange messages, Eve could use her information about the key to try to crack the encryption. She might be successful because even a small amount of information about a key is often enough to detect a pattern in a message if there is one. For instance, if Eve decrypts the message and finds, "I am go**g t* t*e s*p*rma*k**," it is pretty certain she can guess the missing letters.

Another difficulty that Alice and Bob face is that Eve's intercepting and resending photons will have introduced some wrong bit values into the retained bits that make up Bob's key. If this faulty key is then used to decrypt a message sent from Alice, these errors might result in Bob receiving an incorrect message.

Let's say Alice and Bob detect that an eavesdropper is probably tampering with their key distribution, but that only about ten percent or fewer of the key bits might be known to Eve. To eliminate the resulting errors, and to protect against the security risks introduced by the eavesdropper, there is something Alice and Bob can do. They can use mathematical methods called error correction and privacy amplification. These methods rely on Alice and Bob sending more photons than they ideally need to create a key of a certain length.

For example, if they want to create a key with 5000 bits, they might need to send 30,000 photons. This creates a key list of about 15,000 bits (the other half of the bits being discarded because they occurred in cases when Alice and Bob used different polarization schemes). Of the 15,000 bits, some of those in Bob's list don't agree with the corresponding bits in Alice's list. To make a long story short, Alice and Bob can combine certain groups of these bits (say, by adding the values of groups of three) to create a smaller set (say, 5000) of shared bits. The mathematics of probability theory proves that, using this method, nearly all the errors can be eliminated, and Eve ends up with far less information about the new set of key bits than she had about the original set.

With this method, Alice and Bob can make their shared key as error free and as secure as they wish; the price they pay is that Alice must send more photons than the number of key bits in the final key. Sending extra photons is a small price to pay for the benefit of creating, from afar, lists of secure key bits that can be used subsequently for sending nearly perfectly secure messages.

This conclusion about the security of the quantum method for key distribution differs from arguments for the security of 'classical' encryption methods. Using such classical systems, one never knows if a clever adversary has worked out a mathematical method to crack a secret, encrypted message. Remember the Enigma. In contrast, the security of the quantum encryption system rests on the fundamental quantum properties of light, not on the hoped-for limited mathematical ingenuity of the adversary.

Could Eve devise other, better eavesdropping schemes?

You might wonder if Eve could use a better strategy than simply measuring the photons. Perhaps she could make a copy of each incoming photon and measure only the copy. No; recall that it is not physically possible to copy the state of a photon onto another photon. The quantum copier machine is permanently out of order.

In fact, quantum theorists have proved that, according to quantum theory, the nature of light prevents Eve from devising *any* strategy to crack the quantum key sharing method. Only a discovery that quantum theory itself is faulty would invalidate the security of the quantum key sharing method.

What is the current status of quantum key distribution?

Quantum key distribution has proved to be a practical, even commercial, method of sharing secret encryption keys remotely. Secret bits have been transmitted successfully at rates of a million per second over 20 kilometers of optical fiber. Secret keys have been transmitted over fibers as long as 300 kilometers, and ground-to-satellite key distribution is in development. At least four companies sell commercial systems for quantum key distribution. And such systems have already been used in trials for bank transfers and to transmit ballot results.

It seems that quantum technology is marching into the future.

Further Reading

"Quantum Key Distribution," https://en.wikipedia.org/wiki/Quantum_key_distribution.

Simon Singh, *The Code Book: The Science of Secrecy from Ancient Egypt to Quantum Cryptography*. New York, NY: Anchor, 2011.

4

QUANTUM BEHAVIOR AND
ITS DESCRIPTION

How do quantum objects behave in the absence of measurement?

In the previous chapters we discussed the origins of three quantum principles: fundamental randomness, quantum states, and measurement complementarity. The main lesson we learned was that quantum measurement is considered something that either does or does not occur, and it depends very much on which measurement scheme is set up to study the phenomenon. In this chapter, we consider what happens during times when no measurements take place. During these intervals, there are no measurement outcomes that can be recorded or registered, but still a quantum object can move from place to place. I refer to what happens during these intervals as 'quantum behavior.' We will see that such behavior is very different than our common classical intuition would have us believe.

How do electrons and pinballs behave differently?

Imagine playing a pinball game on a machine such as the one shown in FIGURE 4.1. The ball-launching mechanism is adjusted so that every time you launch a ball it lands as nearly as possible at the exact center of the uppermost peg.

You can preset this peg in one of three positions: slanted to the right, as shown in case A; slanted to the left, as shown in case B; or perfectly horizontal, as shown in case C. The two intermediate pegs are fixed. The lower peg is always perfectly horizontal.

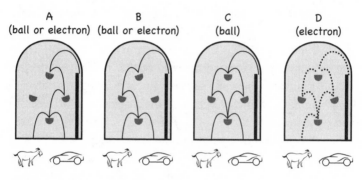

Figure 4.1 Pinball machines for balls or electrons.

In case A, the ball always bounces first to the right, and then at the lowest peg it has an equal chance of bouncing left or right. Depending on which way it bounces, you win either a car or a goat. In case B, the ball always bounces first to the left, and again at the lowest peg it has an equal chance to bounce left or right. So, in these two cases, you have a fifty-percent probability to win either the car or the goat.

In case C, the uppermost peg is horizontal, so when it hits this peg, the ball has a fifty-percent chance of bouncing either left or right. When it arrives at the lowest peg, it again has a fifty-percent chance to go either way. Evidently, in this case there is a fifty-percent probability to win either the car or the goat, just as in cases A and B.

Now consider doing this same experiment with a single electron instead of with a ball. Recall that electrons are particles that have mass, so they fall under gravity similarly to any other object with mass, such as a pinball. Also, electrons are negative electrically charged, so they repel other electrons. Therefore, to construct a 'peg' that causes an electron to bounce from it, we would use a small metal plate on which we have deposited an excess of electrons. A falling electron approaching this surface does not actually touch it but is repelled by it and appears to bounce.

Let's image we conduct this experiment in a perfectly dark room so we can't see the electron as it travels and bounces. For cases A and B, in FIGURE 4.1, the electron behaves similarly to the ball; it may go toward the goat or the car. But, if the electron experiences the same setup as in case C, as shown in case D, an amazing thing happens; at the upper peg, the electron can bounce either way with fifty-percent probability. But then, when the electron bounces from the lowest peg, it always goes left, toward the goat! The probability to win the goat equals one and the probability to win the car is zero. (Sorry about that, if you were hoping for the car.) To emphasize that the quantum pinball behavior is different from the classical physics pinball behavior, I used dots in case D, in FIGURE 4.1, to indicate the two paths the electron could possibly follow. Next I discuss in detail the importance of carrying out the electron experiment in the dark.

Why does the electron always go toward the goat?

The electron always ends up at the goat in case D, whereas the ball in the similar case C goes with fifty-percent probability toward the goat or the car. The fundamental difference in behavior between pinballs and electrons means we must use different rules when calculating probabilities for them.

For case C with the ball, we can calculate the probability to win the goat as follows: The probability that the ball bounces left at the first peg is one-half, and the probability to bounce left at the last peg is again one-half, so the probability to do both is one-half of one-half, or one-quarter—that is, $(\frac{1}{2}) \times (\frac{1}{2}) = \frac{1}{4}$. The same argument holds for the other three possibilities: left/right,

right/right, and right/left. That is, there are four possible paths, all with probability equal to one-quarter. To find the probability of winning the goat, *add* the probabilities for the two distinct paths on the ball that lead to the goat: ¼ + ¼ = ½. The probability of winning the car is found using a similar argument, leading to ¼ + ¼ = ½. The fact that the probabilities of winning the goat or the car are equal is not surprising given the symmetry of the peg arrangement. Even if the pegs were to be moved by tiny distances, the probabilities would still be very nearly equal. It seems the probabilities couldn't possibly be anything other than fifty percent in this situation.

For case D with the electron, there is no obvious way you could combine the probabilities for the two paths leading to the goat (each equaling one-quarter) to create a probability equal to one to win the goat. It must be that considering separate probabilities for separately predictable paths is not the correct way to analyze the problem for quantum objects such as electrons.

We haven't answered the question: Why does the electron always go toward the goat? We need to explore further.

What happens if we modify the setup?

When baffled by the results of experiments, the natural thing to do is to "tweak" the experiment in some ways and see what happens. This might lead us to a new logical rule for describing the situation.

Because it's not clear from the previous cases what is going on, let's try something; let's move one of the pegs. For cases A through C with balls, if we move one of the pegs a tiny amount, it doesn't change the probabilities significantly. In contrast, with electrons traveling in complete darkness, moving one of the pegs slightly can make a very large change in probabilities. In case E of FIGURE 4.2, the rightmost peg has been moved up a tiny distance. For a particular value of this distance, the result is drastic; now the electron always ends up at the location marked by the car! Perhaps even stranger, if we were to move the peg up even farther, doubling the distance moved, as in case F in FIGURE 4.2, we would find that the electron again ends up always at the goat. There is nothing like this in the case of the pinball, which is not sensitive to very small changes of the machine's setup.

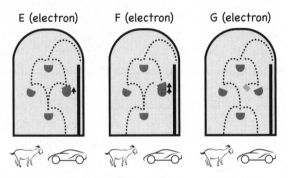

Figure 4.2 Moving a peg or blocking a possible path.

What if we block one path?

Let's try another change in the setup. Let's insert a brick or other barrier that blocks one of the possible paths, as shown in case G of FIGURE 4.2. Now, after several tries of launching the electron, it is found that it goes to either the goat or the car with equal probabilities: twenty-five percent each. When the electron bounces right (fifty percent of the time), it is blocked by the brick. When it bounces left, it is not blocked and is then equally likely to bounce left or right from the lower peg. Compare case G with case D, in which both left and right paths are possible. The change in outcomes if we switch from case G back to case D (by removing the block) suggests that the mere accessibility of the right-going path leading to the car somehow interferes with the left-going path leading to the car, and in case D makes the probability of ending up at the car zero. Again, there is nothing like this in the case of pinballs; adding another possible path for the balls only increases the probability of winning the car, not decreases it.

What can we conclude so far?

From the experiments so far, we can state a tentative rule or principle for quantum probability:

> **Tentative Guiding Principle** If there are two distinct paths an electron can take to arrive at the same location, the possibilities for these paths combine to give a probability that can be greater than or less than the probability predicted by simply adding the probabilities of the paths.

In such cases, it is said the possibilities of two paths 'interfere' with each other, and we refer to this behavior as *quantum path interference*.

This conclusion seems odd, but it has been arrived at through careful experimentation and theoretical modeling. This is not to say we understand it at an intuitive level. It's a fundamental aspect of the nature of electrons.

There is an important aspect of this situation. If both paths after the first peg are open (i.e., not case G), after an electron ends up at either the goat or the car there is no way for us to know that it took either the left-going or the right-going path after the first peg. We ensured the pegs were fixed rigidly in place and the experiment was conducted in perfect darkness so we couldn't monitor the electron's location as it traveled. Maybe we should try changing the setup so we can detect which way the electron travels.

Can we measure which way the electron travels?

Yes, we can—for example, by mounting both intermediate "pegs" on stiff metal springs. Then, if the electron bounces from, say, the right peg, this peg would start vibrating on its spring. The vibrating spring will experience some friction at its mounting points. The friction will heat the metal, and the spring will cease vibrating after all its energy is lost to heat energy. By observing which of the two springs ends up slightly heated after an electron passes through the pinball machine, we can measure which path—left or right—the electron takes between the first and last pegs.

Now we find an interesting result. In the case in which we observe the springs and determine which path the electron takes, the probabilities for the electron to go to the goat or the car turn out to be fifty percent each, just as in the classical physics case with the ball! Somehow our intervention using springs does more than just change our ability to monitor the electron's progress through the machine; it also changes the probabilities of the possible outcomes. This is reminiscent of case G, in which inserting a block into one path removed the phenomenon of interference, making each final outcome equally likely.

It's not our conscious act of knowing which spring becomes heated that changes the probabilities from one and zero to one-half and one-half. Rather, it is the fact that one peg or the other (but not both) is made to vibrate in a detectable manner. The changes in probabilities result from the fact that the electron hitting one of the pegs leaves a *permanent trace*. This permanent trace—the increased temperature of the spring—records information about which path the electron actually took. The recording of a permanent trace constitutes a measurement.

Apparently, then, if a measurement is made on the electron during its travel (even if no one views the measurement result), then the probabilities become what we would predict for a classical object such as a ball. As Alice (in Wonderland) said: "Curiouser and curiouser!"

It seems we need to modify our classical physics concept of 'path' when describing the motion of quantum objects. In classical physics, we think of an object as taking a definite path along some trajectory. In quantum physics, the concept of path splits into the two concepts of *possibility* and detected *outcome*. That is, the starting point and endpoint may be observed or measured as definite outcomes, whereas the behavior in between can be only a set of *quantum possibilities*.

Why can't we apply this same reasoning to the pinball?

Can't we do the experiment with an ordinary ball in a completely dark room, with perfectly rigid pegs, and not have the ability to know which path the ball takes? And then get quantumlike results? No, we can't. For an actual ball weighing, say, a few ounces, it is prohibitively hard to build an experiment such that the ball leaves absolutely no trace of its passing. A ball would create a sound when it hits a peg, and so on. But there is an even more fundamental reason this experiment would be prohibitively hard. It turns out, as we will see later, that the sensitivity to small changes in the setup (as in case E) or to small permanent traces left in the setup (as in the case with heated springs) depends on the weight of the traveling object—that is, how great its mass is. The mass of a pinball is enormously greater than the mass of an electron, so it is enormously more difficult to set up an experiment such as in case D, in which the ball would always go toward the goat as a result of quantum path interference.

This leads to the following commonsense statement: Objects the size and mass of a pinball are never observed to behave like electrons. This realization allows us to distinguish, as a practical matter, a quantum realm of behavior from a classical realm of behavior. Although we know that all matter is governed fundamentally by quantum principles, the observed behavior of objects in general depends on their mass (how heavy they are) and how strongly they interact with their surroundings. It is extremely difficult to make large objects behave coherently in the quantum sense.

What is unitary behavior?

If we set up the electron experiment so the pegs are mounted on springs and a permanent trace of the electron's passing is left behind, then we can divide the electron's behavior into two steps: (1) the electron is launched and goes to the left or right intermediate peg and (2) the electron then goes to the goat or car. There are three recorded observations: the launch, the path in the middle, and the final result. In this situation, our classical physics-based probability calculation holds true; we simply add the probabilities for the possible paths to get the correct probability for the goat or car.

When the pegs are fixed rigidly in place and the room is dark, the electron leaves no trace for us to notice. There are only two recorded observations: the launch and the end result. In this case, we cannot divide its path through the pinball machine into a series of separate steps—each with a definite, observable outcome—as we can do with a ball's path. Because the process of an electron moving through the machine must be considered as a whole, physicists call it 'unitary behavior' or *unitary process*. In English, the word 'unitary' means forming a single entity—having the character of an undivided, whole unit. In physics, unitary behavior is one that cannot be divided into individual observed steps; it takes place wholly 'in the quantum realm,' in which case we can talk only about quantum possibilities, and for which we have no intuitive classical physics explanation or description. A unitary process is coherent, and the 'quantum rules' for calculating probabilities apply to it.

Building on the Guiding Principles stated in Chapter 2, we can add:

> **Guiding Principle #4** There exist in Nature behaviors or processes that are *unitary*, in that they cannot be divided into individual steps, each having a definite, observable outcome. Classical physics fails to recognize such processes.

Recall that, in Chapter 1, we said that electrons, photons, and quarks are *elementary* constituents of matter in that they cannot be divided into yet-smaller constituents. Here we are using the word 'unitary' to refer to physical processes or behaviors that cannot be divided into smaller observed steps. The idea of unitary processes is as close as we can get to complete knowledge of behavior in the quantum world.

What other examples of unitary processes illustrate the main points?

Consider this story: In a totally darkened room, a soccer ball is launched from a known location at a known time. In the room there are a number of fixed objects from which the ball may bounce. Exactly twenty seconds after the launch, the ball arrives in a soccer net located halfway across the room. What can we know or say about the path the ball took between start and finish? We can certainly say it took some particular path (we heard it bounce around), although the path is unknown to us. In the language of measurement used earlier, I call this unknown path an 'unknown outcome.' We are again using the word 'outcome' to refer to some event that actually happened, as opposed to having only the possibility of happening.

Now consider a similar story for a photon of light. A single photon is emitted from an atom at a known time in a totally darkened room. In the room there are a number of shiny, fixed objects from which the photon may bounce, as light bounces from a mirror. Exactly twenty nanoseconds later, the photon is registered by a light detector located halfway across the room. What can we know or say about the path the photon took between start and finish? Here the answer is different from that of the ball. The main difference is that, in the quantum case, the photon can pass through a darkened region and not leave any trace at all—not even a microscopically small trace—that can be used to define the unique path taken by the photon. In such cases, we *cannot* say the photon took a particular, definite path. To claim a photon took a particular path would lead us to make wrong predictions about the probabilities of the photon arriving at detectors at various locations.

The question, "Which path did the photon actually take?" has no answer. The question itself is not valid. It's not just that we don't know which path it took, and it's not that it took neither path nor both paths. Here, we again have the idea of a unitary process.

What are additional consequences of a process being unitary?

Think again about a photon emitted by an atom in a darkened room in which there exist two possible paths from its starting point to the detector. How can we calculate the probability the photon will be registered by a detector at a certain location? To think about such probabilities, consider the experiment in FIGURE 4.3, which shows an atom that emits a photon, and two paths from the atom to an area where there are three detectors, each of which could potentially register a photon. The detector labeled A is positioned on the horizontal centerline, whereas detectors B and C are below the centerline. The photon can be emitted in any direction, at random. A solid (gray) block absorbs the photon if it travels along the horizontal path. Two mirrors are positioned so that a light beam can reflect once and arrive at either detector, depending on the angle at which it departs from the atom. Various possible paths that light could follow are sketched in the figure. If we don't monitor the photon's location, it is not correct to ascribe a probability that a single photon will take any particular path. Still, the paths do exist and are accessible to the photon; they are quantum possibilities. They do affect probabilities for what might ultimately happen—namely, the probabilities that the photon will be registered at detector A or detector B.

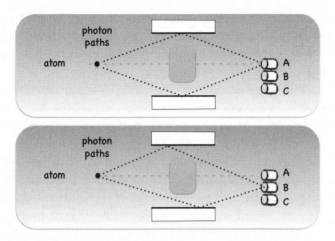

Figure 4.3 A photon may bounce from mirrors as it travels in a darkened room toward detectors.

If, instead of a photon emitted from an atom, let's say we ask the same question about a grain of sand shaken from the fur of a wet dog after a day at the beach. The grain can travel equally likely in any direction, and can bounce from either of the mirrors and arrive at one or the other detector. In this case, we would use the ideas of classical probability theory and note that the probabilities of all possible events will total to one-hundred percent. Let's focus our attention on the cases in which the sand grain reaches detector A. There are two possible ways the grain can reach detector A: by taking the upper or by taking the lower path. We can estimate the probability for detector A success when taking either path and add these probabilities to arrive at the final answer. For example, if we estimate a ten-percent probability for taking the upper path and arriving at detector A, and the same for taking the lower path and arriving at detector A, then the final probability to arrive at detector A is 10% + 10%, which equals 20%, or equivalently we could say the probability equals 0.2.

For the grain of sand ejected by the dog's fur, the probability to reach detector B is roughly equal to the probability of arriving at detector A. This is because the two detectors are close to each other and the grain can be ejected in any direction with equal probability. The same holds for detector C as well.

Now let's return to considering the single photon emitted from an atom, as in FIGURE 4.3. What are the probabilities it will be detected either at detector A or detector B? The answer is different from that of the sand grain. To understand the result, first consider what would happen if we block the lower path. Now there is a ten-percent probability for taking the upper path and arriving at detector A. If we were to block the upper path instead, we would estimate a ten-percent probability for taking the lower path and still arrive at the same detector. So far, this is similar to the case of the sand grain. But now, we will observe a strange result. If we have *both* paths open, there is a *forty-percent* probability for the photon to be detected at detector A! This probability is *not* the sum of the two separate ten-percent probabilities, as it is for the sand grain.

Even stranger, if detectors A and B were separated by a specific distance (which we discuss later in this chapter), there will be a zero-percent probability for the photon to be detected at detector B! That is, we could predict correctly that the photon will never arrive at detector B, although B is very near detector A. Furthermore, we can predict correctly that, in this situation, detector C will not have zero-percent probability to receive the photon. This is another example of the possibilities of two paths seeming to interfere with one another to change radically the final outcome probabilities.

Can matter behave the same as photons in the two-path experiment?

You might think, "Well, a photon is not a constituent of matter, but a constituent of light, which has no mass and carries only a tiny amount of energy. I would be more impressed if matter behaved this way, too. Does it?" Yes. A single electron, which is an elementary constituent of matter, behaves in experiments much like the photon does. In an experiment analogous to the previous photon experiment, but now with electrons, an electron detector labeled B, as in FIGURE 4.3, can have zero-percent probability to register the electron, whereas the two adjacent detectors, A and C, can have high probabilities. This large sensitivity to the details of the experimental setup is analogous to the results we discussed earlier for the 'quantum pinball' machine.

Then you might think, "OK, but a single electron is a tiny elementary object, or perhaps an electron is a 'happening,' rather than an actual object. Maybe, then, it's not so strange that it behaves this way. I would be more impressed if large pieces of matter also behave in this quantum way." Well, they can, but only under special conditions. I said earlier there is nothing in current quantum theory that forbids objects much larger than electrons behaving in quantum mechanical ways. Here is an example:

A carbon nanosphere is a molecule made of sixty carbon atoms arranged in a shape like the seams of an American soccer ball or international football. They are commonly referred to by the nickname buckyballs (after Buckminster Fuller, whose geodesic domes they resemble). The mass of one buckyball is more than a million times greater than the mass of an electron, so it is large by atomic standards. Yet, in experiments similar to the one in FIGURE 4.3, buckyballs have been seen to act just like single photons or electrons. Although a buckyball is not an elementary object, under the right conditions (namely, that it leaves no trace showing a particular path), its travel between two locations is best described as a unitary process. On the other hand, the grain of sand ejected from a dog's fur is far too large and heavy to undergo observable quantum interference behavior, so its motion is best described by classical physics.

Current research is pushing the boundaries of the size of objects that can be made to behave quantum mechanically. For example, the vibratory motion of a tiny glass sphere with a diameter of ten micrometers (containing about sixty trillion atoms) can be observed to exhibit quantum behavior under very special laboratory conditions—very cold and well isolated from its surroundings. Scientists envision technological applications of such research advances.

Can a photon sometimes behave according to classical probability?

This question is the opposite of the previous question, where we asked if a large nonelementary object can behave in a quantum-like manner. Here we ask: Can a single photon behave in a classical physics manner? Here, too, the answer is yes, but only under special conditions. If you'd like to challenge yourself, cover up the following paragraph and try to answer this question: Under what conditions or situations could a photon behave according to classical probability? This would imply that the final probability for the photon to arrive at detector A would equal the sum of the separate probabilities to arrive there by either the upper or lower path (0.2, in the previous example).

You might have figured out that to have the photon behave according to classical probability rules, you need to set up the experiment so the photon leaves a permanent trace that shows which path it took. Then it would be sensible to ask which path it took. Following the example of the quantum pinball machine, we can mount each mirror in FIGURE 4.3 on a spring. The mirrors and spring need to be very lightweight, so a photon can affect them. If the photon bounces from one of the mirrors, its spring will vibrate and heat up slightly from friction, leaving information about which mirror was disturbed by the passing photon. We can say we determined or measured a path for the photon, for which there is now an actual outcome, and we can safely add the corresponding probabilities. This is like the case of the sand grain or the pinball, which behave according to classical physics in the sense that it follows a definite path, even if we don't necessarily know which path that is.

In this case we would observe a 10% + 10%—that is, a 20%—chance for the photon to be registered at detector A. We can check this by repeating the experiment many times and counting the fraction of arrivals at detector A. Even if we do not personally observe which spring was heated, the photon-at-detector-A probability will still be twenty percent.

In this situation, the motion of the photon between its preparation and its detection is *not* a unitary process. It can be divided into an observable sequence: traveling from the atom to one or the other mirror, where an intermediate measurement occurs (personally observed or not), then traveling to the detector. In cases like this, classical probability rules hold true.

How can we summarize the previous considerations as a principle of physics?

From all the previous evidence, we can infer a basic principle of quantum physics:

> **Guiding Principle #5** When a process can be divided into a series of separate steps, each with a definite outcome, we should calculate final probabilities using classical probability (that is, add the two probabilities when two paths lead to the same final outcome). But, when a process cannot be divided into a series of observable intermediate outcomes, then it must be considered 'unitary,' and we cannot calculate final probabilities according to the ideas of classical probabilities.

That is, if given the physical situation we cannot, even in principle, determine a step-wise sequence of intermediate outcomes, then classical probability theory does not hold and we need a new theory of probability. Quantum theory is that new theory.

What is a measurement in quantum physics?

In this and previous chapters I talked a lot about measurements, but perhaps it still isn't entirely clear what quantum physicists mean when they say a measurement has been performed. Using insights we have gained, we can now state the concept more clearly:

> **Guiding Principle #6** A *measurement* is any process that rules out one or more possible outcomes from having occurred. Ideally, it will rule out all but one possible outcome.

As we have seen, placing springs behind small mirrors can result in a permanent trace being left behind if a photon bounces from a particular mirror. This trace will rule out permanently the possibility that the photon bounced from the other mirror. We never see *both* mirrors showing a permanent trace of a single photon. Another example occurs with light polarization. Simply passing a photon through a calcite crystal does not make a quantum measurement. But, if the photon is registered subsequently by a detector, then the crystal and detector work together to create a measurement, the outcome of which has a probability to occur that we can predict using quantum theory.

Can a quantum object exist in two places at once?

No. Although in the absence of intermediate measurements it is wrong to believe a photon takes any particular path, it also is wrong to believe a single photon takes two paths. If we make a measurement, we will detect the photon in just one place. All we can say is that a photon begins in one place and ends up somewhere else. It is more accurate to say the photon does not actually exist at any particular place rather than to say it exists in two places at once.

Quantum theory is not about the trajectories of photons or electrons. This would imply they follow definite trajectories or paths between starting and ending locations. Rather, quantum theory is about probabilities for measurement outcomes at chosen locations.

How does quantum key distribution make use of unitary processes?

In Chapter 3 we saw that quantum key distribution can create excellent privacy because Eve's measurement of a photon disturbs it in a fundamentally quantum mechanical way, allowing Alice and Bob to detect an eavesdropper's presence. Now we can make more clear what we meant by "disturbing something in a fundamentally quantum mechanical way." It means creating a situation in which permanent information about the photon's passing is recorded in the experimental setup, which allows Eve to gain information about the polarization of the photon sent by Alice.

The point is that *any* method Eve devises to gain information about the photon requires a measurement of some kind to be made. As we have seen by the examples in this chapter, such a measurement changes the photon transmission process from a unitary quantum one into a step-wise process—that is, a sequence of outcomes obeying classical probability rules instead of quantum probability rules. This will change the probabilities for observing various outcomes and will thereby allow Alice and Bob to be aware of disturbances caused by Eve.

How does quantum theory describe states in which two possibilities exist?

As we have seen, when an object undergoes a unitary process, there may be two or more possibilities. We have already seen two examples of such situations: (1) a single photon with D-polarization, which simultaneously has H-possibility and V-possibility; and (2) a single photon or electron that travels through a region in which two possible paths exist. In each case, the two possibilities are mutually exclusive, in the sense that if we were to make a measurement in the middle of the process, we would find one or the other possibility, but not both, occurring.

We saw in Chapter 2 how the polarization states of photons can be represented by drawing lines that make up the sides of a right triangle. This was illustrated, for example, in FIGURE 2.9. The line 'points' in the direction of the photon's polarization state. In the language of the current chapter, the longer side of the triangle represents the photon's quantum state, whereas the two shorter sides represent the quantum possibilities. We used Born's Rule, which tells us the square of the length of a side representing a possibility gives the probability to observe the corresponding polarization if we carry out a measurement of the photon.

In fact, this idea of 'pointing' a line in the direction of a quantum possibility applies generally to all quantum systems and to the various types of quantum possibilities: polarization, path, position, energy, and so on.

How does quantum theory describe an electron having two possible paths?

Early on, physicists viewed electrons (constituents of matter) as tiny particles, which presumably should obey the classical physics laws of motion. Instead, let's try applying the ideas just discussed to a case in which the quantum possibilities correspond to two paths available to an electron.

Instead of the artificial example of the quantum pinball machine discussed earlier, consider the setup shown in FIGURE 4.4, in which two paths lead from the electron's starting point on the left to the region in the right side of the figure. Let's label the upper path as path A and the lower path as path B. In the drawings labeled (i) and (ii), there is a detector that can register the passing of the electron, shown as a star, in either the upper or lower path. If a detection occurs in the upper path, we can represent the electron's state as an arrow (↑) pointing to A, which simply means, "The electron was detected in path A." On the other hand, if a detection occurs in path B, as in the drawing labeled (ii), we can represent that state by an arrow (→) pointing to B, which means, "The electron was detected in path B." The state arrows pointing to A and B are drawn perpendicular to indicate they correspond to *mutually exclusive outcomes*. That is, if one happens, the other cannot.

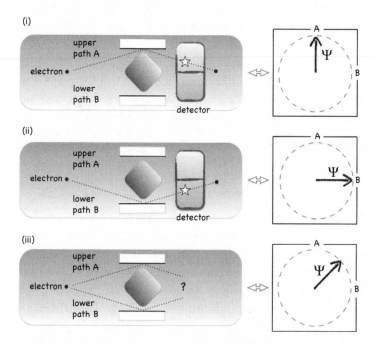

Figure 4.4 (i–iii) An electron leaves the starting location and travels with two path possibilities: upper path A and lower path B. If a detector is present, it registers in which path the electron is detected. An arrow represents the state of the electron by 'pointing' to the measurement outcomes A or B. Part (iii) shows a superposition state of the two possibilities when there is no detector. A circle with a radius equal to one represents a probability equal to one.

The arrow, which I call a *state arrow*, represents the quantum state and is denoted by the Greek letter Ψ, or psi.[1] It is drawn just touching a circle with a radius equal to one, representing that the probability for the corresponding outcome equals one.

In the drawing labeled (iii), both paths are possible for the electron and there is no detector, so these paths interfere with each other in the sense that we discussed earlier in this chapter. In this case, we represent the state as an arrow (↗) that points halfway between A and B. We know that if we were to intervene and detect the electron, we would find an outcome of one path or the other: A or B. In the case shown, the probabilities would be fifty percent for each outcome to occur. But, as long as the electron's behavior is a unitary process (leaving no trace), it is not correct to think that it is in one path or the other, nor can it be said to be in both paths at once.

The type of situation shown in part (iii) of FIGURE 4.4 deserves its own special name. Physicists call it a *quantum superposition state*. The idea is that the A possibility and the B possibility are superimposed on one another in a quantum fashion. There is no counterpart in classical physics for this type of state.

Can arrows be used to represent the state of macroscopic objects?

No, they cannot, typically. Still, it is helpful to consider to what extent this might be possible. For example, FIGURE 4.5 shows state arrows for the states a coin might be in: heads or tails. You can certainly use diagrams such as those in part (i) or part (ii) of the figure to represent the possible states of a coin, but you cannot use a diagram of the type shown in part (iii) to represent the state of any coin, because systems as large and as complex as a coin cannot be said to be in a quantum superposition state. There are no unitary processes available to them.

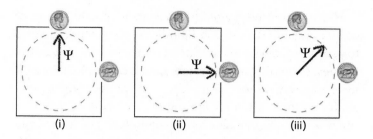

Figure 4.5 (i–iii) Representation of the mutually exclusive states of a coin being 'heads' or being 'tails,' and the hypothetical (but impossible) superposition state of the two.

How are outcome probabilities related to possibilities?

In quantum theory, we combine possibilities rather than add outcome probabilities. For example, the arrow in part (iii) of FIGURE 4.5 can be obtained by adding the arrows in parts (i) and (ii). By "adding arrows," I mean first drawing one arrow and then drawing the second arrow with its tail at the head of the first arrow, and then drawing a new arrow connecting them.

Adding 'possibility arrows' is illustrated in FIGURE 4.6. In this drawing, A and B represent any two mutually exclusive measurement outcomes.

In part (i) of the figure, the state arrow Ψ is made up of equal possibilities for A and B, indicating equal probabilities to observe each outcome if a measurement occurs. In part (ii), a different state arrow Ψ is made up of a larger possibility for the A arrow than for the B arrow, indicating a larger probability to observe the A outcome in a measurement. In our discussion of the Pythagorean Theorem for right triangles in Chapter 2, we saw how Born's Rule is used to calculate outcome probabilities: the length of the Ψ arrow equals one, the lengths of the possibility arrows are a and b, the probability for outcome A equals a^2, and the probability for outcome B equals b^2. The Pythagorean Theorem ensures the probabilities add to one.

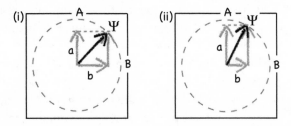

Figure 4.6 (i, ii) Superposition states represented by state arrows pointing in between the A and B outcomes. The lengths a and b of the 'possibility arrows' determine the probabilities for each outcome.

How can an electron be split into two possible paths?

Electrons, if moving at high speed, can pass through thin pieces of matter. If the matter is a crystal, which is made of a regular, repeating pattern of atom positions, then an electron has a certain probability either to pass through in a straight path or to be deflected at a particular, fixed angle, which is determined by the internal structure of the crystal (silicon, for example). Therefore, there are two possible paths by which the electron can emerge from the crystal, as shown in FIGURE 4.7.

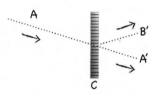

Figure 4.7 An electron approaches crystal C, which has a regular internal arrangement of atoms depicted by a pattern of lines. Whether the electron will travel straight through or be deflected by the crystal is a quantum random event.

How are state arrows used to find probabilities when path interference occurs?

Consider the experiment shown in FIGURE 4.8, in which an electron leaves the starting location and travels with two path possibilities. The two paths converge on a spot on a crystal, at which point the electron can either pass straight through or be deflected. In addition, a path detector may be placed in the region in front of the crystal.

When an electron that has already been detected in path A passes through the crystal, it has equal chances of being detected in the A′ path or the B′ path, as shown in part (i) of FIGURE 4.8. It has fifty-percent probability of each outcome. This probability can be confirmed by repeating the experiment many times. To represent this situation, we draw the state arrow A pointing halfway between the A' and the B' directions.

Part (ii) shows a similar situation, but the B path is the one in which the electron was detected before reaching the crystal. In this case, the state arrow B points in the antidiagonal direction, and it also corresponds to fifty-percent probability for subsequent detection in each of the A' or B' paths. (Notice that the A state arrow and the B state arrow must be at a right angle to each other, because they are mutually exclusive.)[2] In part (iii) of FIGURE 4.8, no path detector is present, and both paths are possibilities for the electron. The A and B possibilities yield a state arrow pointing in the A' direction, and the electron will be detected as the A' outcome all of the time.

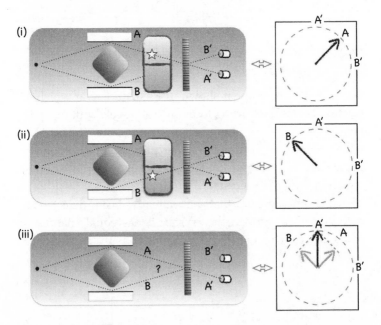

Figure 4.8 (i–iii) The electron has one or two paths possible for it to reach a spot on a crystal. When the electron emerges from the crystal, the probabilities for it to be detected in path A' or path B' may involve quantum interference, as in part (iii).

This result is much like the quantum pinball example we saw at the start of this chapter, where the electron always ended up going toward the goat, as illustrated in case D in FIGURE 4.1. There, I remarked that there was no obvious way you could combine the probabilities (both equaling one-quarter) for the two distinct paths to arrive at the observed goat-winning probability equal to one. And I said we would need to invent new mathematical rules to describe such a result. Here, we have done precisely that! By combining or adding possibility arrows, rather than adding probabilities as you would in classical theory, we have a way to describe how probability works for quantum objects.

What happens if we alter one of the paths?

Do you recall that when discussing the quantum pinball example, I said that moving one of the pegs a tiny distance could change the outgoing path from the goat to the car? In FIGURE 4.8(iii), this would correspond to making one of the paths slightly longer than the other by moving one of the reflecting surfaces. Let's say we move the lower surface down ever so slightly, making the B path a little longer. This slight change can have the effect of flipping the B arrow to the opposite direction, labeled –B (that is, *minus* B) in FIGURE 4.9. When we add the possibility arrows, the resulting arrow points in the B' direction, and the electron would be detected in the B' path.

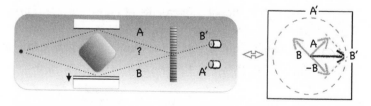

Figure 4.9 Electron interferometer. The lower path is made slightly longer by moving the lower reflecting surface a small distance, and the electron switches to the B' path after passing through the deflecting crystal.

This behavior is an example of path interference, and the setup is called an *electron interferometer*. The method of adding arrows represents nicely the interference of paths and the consequences it has on probabilities.

How can we summarize the previous ideas in a Guiding Principle?

We can restate Guiding Principle #4, which first appeared near the start of this chapter, in a more precise form:

> **Guiding Principle #4 (Precise Version)** If there are two distinct paths (or ways) by which an electron (or any other quantum entity) can arrive at the same final state, the possibilities for these paths are superposed by combining or adding arrows representing the possibilities of each outcome, resulting in the state arrow. The state arrow can be used to find the probabilities for outcomes in any measurement scheme of your choice.

What if we change the path length even more?

For example, say the electron is moving through the apparatus with a speed equal to one-hundred meters per second (which is actually very slow for an electron). Then, the needed change of path length to make the interference switch from the A′ outcome to the B′ outcome is found in experiments to be about three micrometers. For comparison, a human hair has thickness about one-hundred micrometers. Three micrometers is a very small distance—meaning the experiment would be extremely sensitive to small, unintentional movements of either surface. This illustrates why performing such experiments is quite difficult, and perhaps explains why such quantum interference effects were not discovered until the first half of the twentieth century.

Curiously, if we lengthen the lower path by another three micrometers, the electron switches back to the A' outcome! Then, lengthening it by yet another three micrometers, it switches again to the B' outcome, and so on for many cycles, as illustrated in FIGURE 4.10. In this case, six micrometers is the distance required to cycle back and forth once. Let's call this required distance the *full-cycle length*. Such an effect occurs with either electrons or photons. The details are a little different, but let's not get distracted by that.

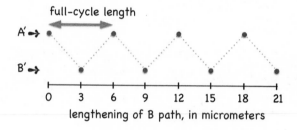

Figure 4.10 Lengthening path B in the electron interferometer in steps of three micrometers causes the output to switch periodically between the A' outcome and the B' outcome.

The picture in FIGURE 4.10 is similar to a ruler—a device with regularly spaced tick marks that we often use to measure distance. We can imagine that the electron carries with it a 'quantum ruler' that it uses to measure the lengthening in either path. FIGURE 4.11 shows such a ruler in each path. The ruler has major tick marks separated by one full-cycle length, and minor tick marks halfway between. Each ruler measures the relevant length of a path from the starting location on the left, although only a portion of each ruler is drawn in the figure.

The important feature to notice in each case is how the rulers line up when they intersect at the spot on the crystal. Let's first look at case (i). At the crystal, the rulers are aligned with like tick marks meeting up (major-to-major and minor-to-minor). The electron is detected in the A′ path. In case (ii), the lower path has been lengthened, resulting in the ruler in the B path being pulled back by one-half of a full-cycle length (three micrometers in our example), so now it is 'behind' the A ruler by one-half of the full-cycle length. Now the major tick marks line up with the minor ones. The electron is detected in the B′ path. As we increase the B-path delay further, the detected outcome will flip back and forth between A′ and B′ as shown in the graph in FIGURE 4.10.

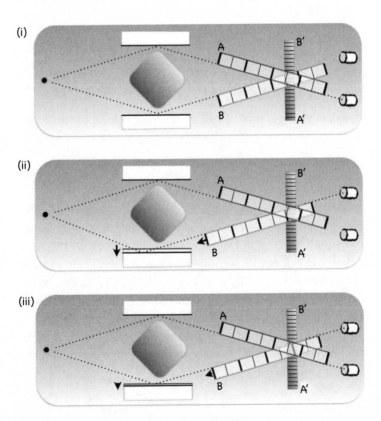

Figure 4.11 (i–iii) The 'quantum ruler' picture for visualizing the interference of paths.

What happens if we lengthen the B path gradually instead of in jumps, as in the previous example? We observe that, for certain path lengths, there is no certainty about which path the particle will be detected in at the output. Each outcome can only be said to be more or less probable. For example, for a B-path lengthening of 1.5 micrometers or one-quarter of a full-cycle length, as in FIGURE 4.11(iii), there is a fifty-percent chance the outcome will be A' and a fifty-percent chance the outcome will be B'.

In FIGURE 4.12, the probability for the A' outcome is shown as a smoothly changing curve as the path is lengthened. The probability for the B' outcome appears as a similar curve, but it is shifted to the right, so when the A' probability is one, the B' probability is zero. For any particular value of the path lengthening, these add to one, as probabilities must.

The closer the rulers are to matching their like tick marks (major–major or minor–minor) at the crystal, the greater the probability is that the outcome will be the A' path. The closer

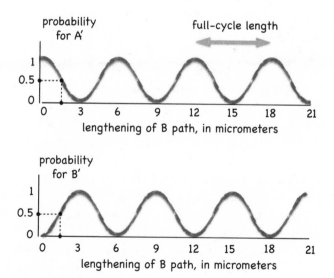

Figure 4.12 Lengthening path B in the electron interferometer causes the probabilities for the A' and B' outcomes to vary smoothly between one and zero.

the rulers are to matching opposite-type tick marks (major–minor), the smaller the probability is that the outcome will be the A' path.

Is there a general principle we can infer from this experiment?

Although this fictitious ruler comparing scheme for understanding quantum interference might seem silly, it corresponds exactly to the mathematics used in quantum theory to describe a very large number of experiments. To capture this idea, I make the following statement:

> Every elementary particle acts like it carries a *quantum ruler*.

The nature of this imaginary or fictitious ruler depends on the existence of a previously unrecognized fundamental constant of Nature called Planck's constant. It is denoted by the letter h, and was discovered by Max Planck in 1900. Planck's constant has units of time multiplied by energy. In units in which time is measured in seconds and energy is measured in joules,[3] its value is exceedingly small: $h = 6.6 \times 10^{-34}$—a number less than one divided by a billion trillion trillion. Why so small? Because it has meaning only on the scale of single electrons or single photons, and these carry very small amounts of energy indeed.

When experiments are carried out with the electron interferometer for different speeds of the electron, it is found that to make the interference switch from the A' outcome to the B' outcome, the B path needs to be increased by a length equal to one-half of Planck's constant divided by the product of the electron's speed multiplied by its mass.[4] FIGURE 4.13 shows examples of quantum rulers for an electron having two different speeds. The full-cycle length is indicated by adjacent tick marks, which are separated by a distance equal to Planck's constant divided by the product of the electron's speed multiplied by its mass. The electron on the right has a speed fifty percent faster than the electron on the left, and so has fifty percent more full cycles (six vs. four) in the same length of ruler than the ruler for the slower electron.

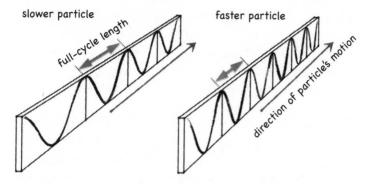

Figure 4.13 Quantum rulers associated with a slower or faster electron.

Physicists use the word *momentum* for the speed of a particle multiplied by its mass. So we can say the major tick marks are separated by Planck's constant, *h*, divided by the particle's momentum:

$$Distance\ between\ major\ tick\ marks\ on\ the\ quantum$$
$$ruler\,(full\text{-}cycle\ length) = \frac{h}{Momentum}.$$

This is called de Broglie's **length–momentum relation**, and it was first hypothesized by a French PhD student, Prince Louis de Broglie, in 1924.[5] This idea was unexpected, and it was visionary in that de Broglie hypothesized it before any electron interferometer experiments had been performed. Five years later, after its experimental confirmation by other physicists who observed electron interference, de Broglie was awarded a Nobel Prize.

What are the take-away messages from this chapter?

In this chapter we discussed some of the basic facts and ideas that underlie the origins of quantum theory. The view of the physical world that quantum physics presents to us is radically different from that of ordinary experience. The latter is captured well by the ideas of the classical physics of Newton, which uses the commonsense ideas of particles, trajectories, and classical probability. Quantum physics introduces a fundamentally different way to describe Nature at its most elementary level. The main new idea introduced by quantum theory is that of the quantum state. This alternative description is necessitated by experimental facts.

The main difference between the classical description of Nature and the quantum description is this: In the classical description, the idea of 'state' is directly equivalent to the set of measurement outcomes, because the state is presumed to specify those outcomes with complete certainty. In contrast, in a quantum description, the idea of state is equivalent to the set of *probabilities* for all possible measurement outcomes, which are fundamentally random or probabilistic.

A quantum state can be represented by an arrow that 'points' to different possibilities for measurement outcomes. We found, by considering experiments, that for quantum entities, probabilities do not add when there are two ways to arrive at the same outcome. Instead, when there are two possible ways a system could yield the same outcome, the corresponding state arrows combine or add, and the new, resulting state arrow tells us, through Born's Rule, the probability of any outcome. This phenomenon is called interference.

A key to understanding quantum behavior is the idea of a unitary process—one that cannot be divided into a sequence of observable outcomes. This is the case if there is nothing in the process that leaves a permanent mark or trace of the quantum object's passing. On the other hand, if permanent traces are left, then we say a measurement occurred; then the classical theory of probability does apply and probabilities do add.

We found that in an interference experiment with electrons, lengthening one path affects the interference of the arrows representing the two possible outcomes. The full-cycle length that determines the outcome probabilities depends on Planck's constant, h, which takes its honorary place beside other fundamental constants of Nature, such as the speed of light and Newton's gravitational constant.

At this point, you might be wondering what causes the curious behavior of the A' and B' outcomes when one path is lengthened. And well you should! This kind of result certainly has no counterpart in the classical physics of particles, so it reinforces that electrons cannot be considered to be ordinary particles. The fact that the outcome changes periodically between two values turns out to be a big clue for understanding the structure of atoms, in which electrons are trapped in a small region surrounding an atomic nucleus. We explore this topic in Chapters 6 and 13.

Notes

1   'Psi' is pronounced like the name of the famous Korean pop star Psy, where 'ps' sounds as in the word 'psychology'; in other words, the 'p' is silent.

2   Because parts (i) and (ii) in FIGURE 4.8 look similar, you might wonder why A points diagonally and B points antidiagonally, and not vice versa. Actually, we can draw it that way and still reach the same conclusions. The point is that both states must correspond to fifty-percent probabilities for A' and B', and they must be represented by state arrows that are perpendicular.

3   One joule equals the energy emitted by a one-watt light bulb in one second.

4   To write this as a formula, denote the electron's mass by M (about 10^{-30} kilograms) and its speed by S. Then, the needed change of path length is $(1/2)h/(M \times S)$.

5   de Broglie conceived of this relation as representing a wave phenomenon, but I have avoided introducing the wave concept until Chapter 6, so you don't get the idea a wave in this context is something 'physical' like a classical wave.

5

APPLICATION: SENSING GRAVITY WITH QUANTUM INTERFERENCE

What is the technology of sensing?

The human body has at least five senses: touch, taste, smell, sight, and hearing. A human sense is any biological process that generates a noticeable signal informing a person of a physical stimulus or condition. Each involves a bit of physics and/or chemistry. For example, when we touch a warm or cold object, the nerves in our skin can sense the warmth or coldness of the object. A more subtle sense that you have is the ability to feel acceleration when, for example, an elevator in which you are a passenger suddenly starts moving upward. The muscles in your body feel an extra strain or tension, and nerves communicate to your brain that you feel acceleration. The same sense is used to feel the strength of gravity, which, for example, is weaker on the Moon than on Earth.

Humans have created technology that outperforms the human senses. A sensor is any physical or chemical process that generates a measurable signal informing of a physical stimulus or condition. A telescope can 'see' clearly at much greater distances than the unaided eye. A strain gauge can detect movements as small as a millionth of an inch. Factory robots use sensing technologies to see and feel the objects they are manipulating. Roving vehicles in nuclear contamination sites have radiation sensors. Rovers on Mars are equipped with sophisticated sensors for detecting chemical compounds such as water and minerals, as well as organic chemicals, which might indicate the presence of life.

Scientific experiments all rely on some form of sensing. In physics, experimental results are stated quantitatively—that is, in terms of numbers: How far? How fast? How heavy? How hot? In fact, the history of physics is tied to the progress in human-made instrumentation, which allowed ever-more accurate measurements of physical phenomena: the telescope; the microscope; methods for measuring distance, time, mass, temperature, amount of electric charge, and magnetic field strength; and so on. Many of these can now be measured to a precision of one part in a billion or better.

Why is sensing the strength of gravity useful?

A device that can sense gravity's strength can be used for mapping the varying strength of gravity over a geographic area. For example, say an area contains a deposit of gold below it, whereas an adjacent area has only ordinary earth below it. Because gold is very dense, the strength of gravity above the gold deposit will be slightly greater than the strength of gravity in the other area. I think you can see easily the usefulness of measuring gravity in this case.

A more practical application is detecting oil deposits underground, or even mapping archaeological structures under the ground—without digging! An especially interesting application is mapping long-forgotten sewer pipes or old subway tunnels under a busy city, where you really don't want to be digging big holes for exploration.

How can quantum physics be used to sense gravity?

The first experiment for detecting the effects of gravity in a quantum mechanical setting was carried out using neutrons. Recall, neutrons are electrically neutral particles that normally exist in the nucleus of atoms. When released from the atomic nucleus by a nuclear reaction, they can be entrained into a directed 'beam' moving in one direction inside an experimental apparatus.

FIGURE 5.1 shows the interferometer for neutrons that was constructed by Roberto Colella, Al Overhauser, and Sam Werner in 1975. The left side of FIGURE 5.1 is a drawing taken from their research publication. The right side FIGURE 5.1 is a side view. The interferometer was carved out of a single three-inch-long crystal of silicon and has three parallel silicon plates through which neutrons can pass. Neutrons pass easily through matter because they are not subject to electric forces. Each neutron has a probability either to pass through in a straight path or to be deflected at a particular angle, which is determined by the crystal structure of the silicon. Whether a neutron will travel straight through or be deflected at each of the crystal surfaces—labeled A, B, C, and D—is a quantum random event. Therefore, there are two possible paths by which the neutron can reach detector C_2, and the same holds for detector C_3. The physics behind this device is essentially the same as for the electron interferometer described in Chapter 4.

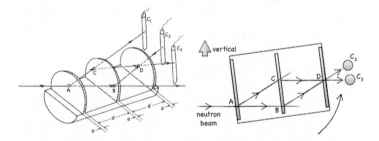

Figure 5.1 Perspective drawing of the neutron interferometer and the side view of the same. A neutron enters from the left and has probabilities of arriving at either of the neutron detectors C_2 or C_3. Quantum interference determines these probabilities. (The left panel is used with permission of the American Physical Society.)

The motion of a neutron between location A and a detector is a unitary quantum process and cannot be broken into separate observable steps. Therefore, we expect quantum interference to influence the probabilities of detecting the neutron at either detector.

To see how gravity affects the interference, recall that de Broglie's relation is the relation between a particle's momentum (speed multiplied by mass) and the spacing of the marks on its 'quantum ruler.' It is expressed by

$$\text{Distance between major tick marks on the quantum ruler} = \frac{h}{\text{Momentum}},$$

where h is Planck's constant, as before.

Now we can see how gravity affects the interference. When the neutron comes into the interferometer with a certain speed—that is, a certain momentum—it has an associated quantum ruler with a full-cycle length (distance between major tick marks) determined by its momentum. If the neutron were to deflect and climb up against the force of gravity along the A-to-C path, it would slow down (like a snowboarder sliding up the side of a half pipe). Then, as it travels along the C-to-D path, it would move slower than if it had remained on the lower A-to-B path. This means the quantum ruler has a longer full-cycle length in the upper path than in the lower path.

Therefore, when the paths meet up at location D, the quantum possibilities combine or interfere in a manner that depends on how much slower the neutron travels in the upper path than it does in the lower path. This in turn depends on how high the particle climbs in the vertical direction along the A-to-C path.

The researchers found that if they gradually tilted the whole apparatus counterclockwise, as shown by the curved arrow, so the neutron had to climb higher and higher to reach the C-to-D path, then the neutron beam was observed to switch gradually back and forth between the C_2 and C_3 detectors.

This behavior is illustrated in FIGURE 5.2, which shows the apparatus with two different tilt angles, and the quantum rulers for each case. The essential difference (not evident in the figure) is that the scales on the rulers are different for portions of the two paths. The distance between tick marks is longer on the ruler for the neutron when traveling along the C-to-D path than when traveling on the A-to-B path because it is traveling slower. In the left side of FIGURE 5.2, the possible paths interfere such that the major tick marks line up with each other at location D, and the neutron has the highest probability to be detected at C_3. In the right side of FIGURE 5.2, the possible paths interfere such that the major tick marks in the upper path line up with the minor tick marks in the lower path, and the neutron has the highest probability to be detected at C_2. As the tilt of the apparatus is increased further, detector C_3 would again have the higher probability to register neutron detection, and so on.

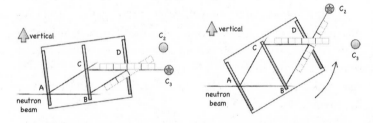

Figure 5.2 When the neutron interferometer is tilted so the upper path is raised higher, the effect of gravity influences the interference and causes the emerging neutrons to switch from detector C_3 to detector C_2.

How rapidly the outcome switches back and forth between detectors as the tilt angle changes depends on the strength of gravity in the vicinity of the apparatus. Thus, this apparatus could be used to detect differences in gravity's strength by moving it from place to place and repeating the experiment at each location.

How is this interferometer different from the one discussed in the previous chapter?

The interferometer discussed at the end of the previous chapter is not influenced by gravity because, in that case, the two possible paths were in the same horizontal plane—that is, at the same height relative to gravity. The figures in the previous chapter should therefore be viewed as top views rather than side views, as are the figures in this chapter.

Is this apparatus a practical gravity sensor?

No, it is not. A major drawback of the neutron apparatus is that it cannot be made easily portable. The source of neutrons was a radioactive uranium (U-235) fuel rod in a nuclear reactor, which is not easily or safely transportable, to say the least. To be useful as a gravity sensor, the device needs to be portable so it can map the strength of gravity over some geographic area.

Present-day quantum gravity sensors, or gravimeters, are based on atoms, not neutrons. Atoms have internal structure—namely, electrons confined tightly within the space surrounding a nucleus. I explain in Chapter 13 how the quantum behavior of atoms leads to the technology of fantastically precise atomic clocks and gravimeters.

Figure Notes

The left panel of Figure 5.1 is taken, with permission, from R. Colella, A. W. Overhauser, and, S. A. Werner, "Observation of Gravitationally Induced Quantum Interference," *Physical Review Letters* 34 (1975), 1472.

6

QUANTUM POSSIBILITIES AND WAVES

How does the concept of waves enter quantum theory?

In quantum theory, probabilities are determined by considering all the possibilities that may be involved in a certain process. If there are intermediate possibilities involved in a process, quantum theory tells us how to combine them to find the resulting possibilities. From these resulting possibilities, the probabilities for a particular measurement outcome can be determined. Quantum possibilities for objects such as electrons behave in some respect as waves do.

But an electron is a single particle—so what is 'waving'? Usually we think of a wave as a disturbance traveling on some extended physical medium, such as sound waves through air or ripples traveling on a lake's surface. That same physical concept of a wave can't apply to a single electron, which, if detected, is found to be at a point, not spread throughout some region. Nevertheless, the wave concept can be applied to single electrons, because it describes correctly how the quantum possibilities that correspond to different measurement outcomes change in time and vary throughout space. This chapter explores the use of the wave concept in quantum physics.

What are waves?

Waves are coordinated patterns that move through space. If a rock is tossed into a pond, it creates ripples on the water's surface that travel away from the location hit by the rock. The ripple pattern moves across the lake, while the water molecules each oscillate around their own fixed positions. The molecules rise and fall, causing adjacent molecules to rise and fall, time-delayed a little from the motion of the surrounding molecules. This coordinated motion of water molecules leads to energy and momentum being transferred along the surface. A duck floating some distance from the source of the wave (the rock entry point) will be affected by this energy and momentum, and will oscillate up and down. But note that this flow of energy and momentum does not involve the water actually flowing between the source of the wave and a location where its effects are felt (the duck).

An example of a wave is shown as a moving pattern in FIGURE 6.1. Locations of maximum wave height are called 'crests' and locations of minimum wave height are 'troughs.' The pattern between two adjacent crests (or two troughs) is called a *full cycle*. The distance between two adjacent crests is called the *full-cycle length* (also called the 'wavelength'). The pattern or wave appears to move forward, although there are no objects moving along in the direction the wave is traveling.

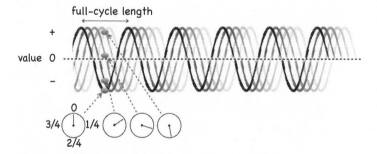

Figure 6.1 A wave is depicted having a value (wave height for a water wave) that oscillates positively then negatively in a regular manner. The distance between crests is the full-cycle length. The time elapsed during one complete oscillation cycle is the full-cycle time.

A point (or a duck) on the wave oscillates up and down once in a characteristic time called the *full-cycle time* (also called the oscillation 'period'). The figure shows a clock that keeps time while a particular point on the wave oscillates up and back down. I call this the 'internal clock.' The rate at which the clock hand rotates around the clock face depends on the springiness of the wave medium, such as water or air, and its density. When the wave's internal clock reads zero, the wave pattern is positioned as shown by the darkest curved line in FIGURE 6.1. As time goes on and the clock hand rotates, the wave pattern moves smoothly to the right, as shown by the sequence of lighter curved lines. Every time the clock hand goes around once, the wave travels a distance equal to the full-cycle length.

The speed at which the wave moves is equal to the length divided by the time. This is the *wave speed*:

$$Wave\ speed = \frac{Full\text{-}cycle\ length}{Full\text{-}cycle\ time}.$$

This relation between full-cycle length, full-cycle time, and wave speed can be visualized in a cartoonlike way with a fictitious ladder, as in FIGURE 6.2.

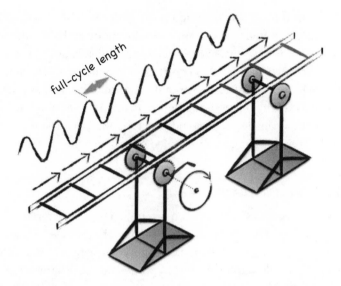

Figure 6.2 A mechanical representation of the relationship between full-cycle length, full-cycle time, and wave speed.

The distance between ladder rungs represents the wave's full-cycle length. The ladder rests on a mechanism that is built to move it using a crank-and-wheel arrangement. The wheels connected to the crankshaft grip the bottom edge of the ladder by friction, so when the crank is turned, the ladder is transported in the direction shown by the arrows. The circumference of the wheels is such that if you turn the crank one time fully around, the ladder moves by a distance equal to the distance between rungs of the ladder. Therefore, if you turn the crank at a steady rate, the ladder is propelled forward continuously at a steady speed. For example, if you turn the crank once every second and the rungs are separated by one foot, then the speed of the ladder is one foot per second.

What is wave interference?

If two rocks are tossed together into a pond, each creates waves in the form of ripples moving away from the location where a rock hit. If a duck is floating on the water nearby, it will feel the effects of both waves. At certain locations in the pond, the actions of the two waves reinforce each other, causing a large up-and-down oscillation. (The duck has a wild ride.) This reinforcing effect is called constructive interference. But, at other locations in the pond, the actions of the two waves may cancel each other, causing no up-and-down oscillation. (The duck is still.) This canceling effect is called destructive interference.

What are quantum possibility waves?

As discussed in Chapter 4, a quantum ruler is a pictorial representation of the fact that every elementary particle has associated with it a characteristic length—a quantum full-cycle length. Recall Louis de Broglie hypothesized this characteristic length as being determined by Planck's constant divided by the particle's momentum. In a two-path experiment, the quantum ruler determines how interference of quantum possibilities influences the probabilities for different outcomes.

de Broglie hypothesized that the tick marks of an electron's quantum ruler move in the same direction the actual electron is moving. The faster the electron, the faster the tick marks move. Curiously, though, the tick marks don't move with the same speed as the electron itself.[1] The motion of the quantum ruler does not represent directly the motion of the electron when viewed as a particle; rather, it keeps track of how quantum possibilities interfere.

de Broglie's hypothesis can be illustrated by again referring to FIGURE 6.2, in which the ladder represents a particle's quantum ruler, and the ladder rungs represent the ruler's major tick marks. The smoothly oscillating curve associated with the quantum ruler is also drawn in the figure, with its full-cycle length equal to the distance between rungs. Imagine that the quantum ruler moves along with the ladder. The moving oscillating curve is a quantum wave, which can be used to determine probabilities for detecting the electron at different locations if a measurement is made.

A moving wave appears to be associated with the electron, but what is 'waving'? Quantum *possibilities* are 'waving.' A **quantum possibility** *wave* does not involve a physical medium. Here we have a highly abstract representation of an aspect of Nature: the fact that quantum possibilities oscillate in space and time in a wavelike manner. Physicists were led to develop this abstract representation by considering experimental observations that couldn't be explained using classical physics theory. The example used in Chapter 4 to make this point was the interference of quantum possibilities that takes placed when a particle has two possible paths available to it as it travels toward a detector. The way in which the quantum possibilities combined and interfered suggested an analogy with wave interference.

Erwin Schrödinger denoted the quantity—quantum possibility—that is waving by the Greek letter ψ, or 'psi.' This is the lowercase version of the uppercase Greek letter Ψ, which I used in Chapter 4 to symbolize state arrows. We can summarize these theoretical developments as follows:

> The psi wave, ψ, represents the time-and-space evolution of the quantum possibilities associated with a quantum object such as an electron.

The psi wave does not represent the electron directly; rather, it represents the information—in the form of quantum possibilities—used to predict outcomes of quantum measurements made on the electron. A mnemonic helps in remembering what the psi wave stands for: psi wave equals PosSibIlity wave.

How does a psi wave keep track of its internal timing?

Because the fictitious crank in the cartoon model (FIGURE 6.2) turns fully in one full-cycle time, it would seem the electron must have some internal mechanism for keeping track of its internal timing. To have a vivid picture in our mind, let's call this timing action of an electron its 'internal clock.' Of course, there is no actual clock inside an electron. An electron's internal clock is merely a way for us to think about how an electron's psi wave evolves in time.

To make our model work, the electron's internal clock must tick regularly, once each full-cycle time. The full-cycle time is the time it takes for the clock's hand to go completely around the clock face once. In an ordinary wall clock, the second hand goes around completely once every minute, so one minute is its full-cycle time.

So, in addition to saying that every elementary particle acts as if it carries a quantum ruler, it also acts as if it carries an internal quantum clock! We can say:

> Every elementary particle has its own timing—that is, it acts as if it carries an internal clock.

It tracks both time and space by itself because of its very nature. The two concepts—quantum clock and quantum ruler—together describe quantum possibility waves.

What sets the cycle time or ticking rate of a particle's internal clock?

In answering this question, de Broglie was inspired by the earlier ideas of Max Planck and Albert Einstein mentioned in Chapter 1. According to Planck and Einstein, the full-cycle time of the quantum wave associated with an elementary particle (electron or photon) is determined by the particle's energy. It is calculated as Planck's constant (denoted h) divided by the particle's energy. We can summarize this relation by the following:

$$\textit{Full-cycle time of internal clock} = \frac{h}{\textit{Energy}}.$$

This is called Planck's *time–energy relation*. It is reminiscent of de Broglie's length–momentum relation that defines the full-cycle length, as discussed in Chapter 4.

The time–energy relation can be expressed in a different way, using the fact that the rate, or frequency, of a clock's ticking is the fractional inverse of the full-cycle time. For example, if the full-cycle time is one-tenth of a second, the frequency is ten cycles per second. So Planck's formula for the ticking rate, or frequency, is written

$$Ticking\ rate\ (frequency)\ of\ internal\ clock = \frac{Energy}{h}.$$

This means the greater the energy of an elementary particle, the faster (more frequently each second) its internal clock ticks. Planck's constant is the constant of proportionality between these two and, as such, is a fundamental constant in Nature.

In Chapter 4, it was stated that if time is measured in seconds and energy is measured in joules, Planck's constant is $h = 6.6 \times 10^{-34}$. This means that if the particle were to have energy equal to 6.6×10^{-34} joules (an exceedingly small amount), its internal clock would tick once per second. This would be very slow for a particle's internal clock. For example, a blue photon has energy equal to about 5×10^{-19} joules. Then, Planck's relation predicts a ticking rate of about 1×10^{15} ticks per second. That is enormously fast! It is this high rate or frequency of internal clock ticking that makes modern atomic clocks extremely precise and therefore of great technological importance, as I describe in Chapter 12.

The energy of an *electron* depends on how fast it is moving. Therefore, electrons with different speeds have different frequencies of ticking of their internal clocks. The energy of a *photon* depends on its color. Therefore, different-color photons have inherently different frequencies of ticking of their internal clocks.

How can we assemble our Guiding Principles into a coherent quantum theory?

So far we have discussed many tools useful in defining the behavior of matter (electrons, neutrons, protons) and light (photons): measurements, probabilities, quantum possibilities, unitary processes, complementary outcomes, quantum states, quantum superposition, path interference, de Broglie's length–momentum relation, and Planck's time–energy relation. It's time to try putting all these ideas together into a coherent picture of the way in which quantum effects move through space.

Imagine you are a scientist in the 1920s facing the task of making sense of all these phenomena and concepts. How would you go about solving the puzzle of quantum physics? Some of the parts were assembled by de Broglie in 1924, who realized that his quantum length–momentum relation, when combined with Planck's time–energy relation, suggested strongly the existence of a previously unknown 'quantum wave' associated with each electron, later to be called the psi wave. Previously, it was thought waves could be associated only with radio signals or with light or with large numbers of vibrating particles. de Broglie's associating a wave with an electron allowed him to predict that electrons would undergo interference effects, as were later observed in experiments.

In 1925, Erwin Schrödinger tried to understand the inner workings of atoms by combining de Broglie's and Planck's ideas on the relations of momentum and energy to length and time. When doing so, he discovered the most important theoretical equation of quantum physics: *Schrödinger's equation*. For inspiration, I showed it in Note 3 in Chapter 1. Its predictions have all been confirmed experimentally with great accuracy, leaving no doubt that it contains the correct physics needed to describe all atomic phenomena. It boggles the mind that such a powerful theoretical formalism can be written in less than a single line. If beauty in science is simplicity combined with power, then this is perhaps the most beautiful of all physics equations.

I have already discussed the essence of what we need to arrive at Schrödinger's equation—namely, according to quantum physics, energy is a quantity that is related to time, and momentum is a quantity related to length. We need to flesh out these ideas and put them together. Before discussing Schrödinger's equation, let's go a little deeper into the meanings of momentum and energy.

What is momentum and what can change it?

Momentum, or inertia, is the idea that objects tend to resist changing their speed and their direction of motion. That is, inertia is the tendency of an object at rest to remain at rest, and the tendency of an object moving in a straight line at constant speed to continue moving in that manner. Physicists use the word 'momentum' to mean the velocity of a particle multiplied by its mass. You can think of momentum as 'that which keeps an object moving in a straight line in the absence of any forces acting on it.'

When two objects collide with each other (as, say, two hockey pucks colliding on ice), some momentum can be transferred from one object to the other, but the total amount of momentum shared between them does not change; it is constant. In fact, the total amount of momentum shared between two interacting objects remains constant. Physicists say that momentum is 'conserved,' by which they mean unchanging in total.

Momentum is, in a sense, intangible; it is an abstract concept. We can't touch it or hold it. Physicists can't really say what momentum *is*, other than by using mathematical equations and by talking about the *effects* of momentum.

To change an individual object's speed or direction requires some force acting on it. *Force* is an abstract concept that refers to the interaction of objects or of fields and objects. In the absence of a force, the object has constant momentum. Long before Schrödinger took up the challenge of quantum physics, Isaac Newton introduced his now-famous equation of classical physics to describe how an object's momentum changes under different conditions. An individual object's momentum can change quickly or slowly, depending on the forces present. For an individual object that has mass (footballs, electrons, and so on), Newton's equation is[2]

> *Rate of change in an object's momentum = Force applied.*

For a particle with mass, momentum is its velocity multiplied by its mass. So Newton's equation means the rate of change of velocity equals the force applied divided by the object's mass. Given the same force applied, a more massive (heavier) object experiences less change of velocity than does a less massive object.

By solving Newton's equation, physicists can predict a classical trajectory for a particle. This trajectory is the classical *state* of the particle, and it predicts the outcome perfectly if a measurement is made of its position or momentum, or both. Recall, however, that in quantum physics, 'trajectory' loses its meaning; we cannot ascribe a definite position or momentum to an object while it undergoes a unitary process of transiting from one place to another.

Light also has momentum, even though it doesn't possess mass. To see this, consider that if light strikes an object, it can 'kick' the object into a new state of motion. For example, if light reflects from a mirror, it kicks the mirror a tiny amount and causes the mirror to recoil, or move slightly, just as a bullet ricocheting from a metal block causes recoil of the block. That is, the light transfers some of its momentum to the object; the light's momentum decreases. But, because light doesn't possess mass, Newton's equation does not apply to light, which is described by a different theory called Maxwell's equations.

What is energy?

Energy is another intangible concept, the general meaning of which is the capability to cause motion. You can think of energy as 'that which keeps a clock ticking in the absence of any significant forces slowing it down.'

Energy is different from momentum, as seen from the following example. The bob in a pendulum clock does not have constant momentum during each of its swings, because each time it travels in an arc, gravity exerts a downward force, bringing it to rest momentarily at its uppermost height. By the bob attaining greater height, the energy that was associated earlier with the motion, or momentum, of the bob has been stored for later use. Then, the stored energy begins being converted back into energy of motion, and the bob gravitates back along its arc-shaped path. This continual exchange between 'stored energy' and 'energy of motion,' and the fact that the combined energy is constant, is what keeps the pendulum swinging and the clock ticking.

The total energy of an object equals the sum of the energy of motion and the stored energy. This idea can be written as an equation. The energy equation is

$$Total\ energy = Energy\ of\ motion + Stored\ energy.$$

Physicists refer to stored energy as potential energy, and energy of motion as kinetic energy.

Energy, like momentum, is conserved—that is, unchanging in total. Energy cannot be created nor can it be destroyed, but it can be exchanged between different objects and between different forms. To see that energy can take different forms, consider the following example. When you eat a candy bar and then later kick a football, some of the chemical energy stored in the candy is converted via your muscles into your leg's energy of motion. Examples like these are the reason scientists recognized energy as being a universal quantity.

Now, with an understanding of energy and momentum in hand, we return to the question of quantum psi waves.

How does Schrödinger's equation describe quantum objects moving through space?

Unlike a classical object such as a football, an electron does not really have a trajectory. Its 'motion' is a unitary process.

As long as it is not being detected and is not leaving a permanent trace of its whereabouts, its motion cannot be broken down conceptually into a series of definite, unique steps. Therefore, we can't use Newton's equation to describe its motion in terms of a trajectory. Instead, Erwin Schrödinger reasoned that we should describe how the *quantum state* of the electron changes in time as a unitary process. He developed an equation for doing so in terms of de Broglie's quantum possibility wave, which represents the state of the electron. That is, it represents the possible outcomes of measuring the electron's position.

Schrödinger's equation describes mathematically how the shape of the quantum possibility wave changes as the associated electron 'travels' over all possible locations. Schrödinger used Planck's ideas on internal quantum clocks and de Broglie's ideas on quantum rulers and waves, combined with the ideas of energy of motion and stored energy, to produce his famous equation for the psi wave, which I represent in words. Schrödinger's equation is

$$\begin{pmatrix} \text{Time rate of change} \\ \text{of psi wave} \end{pmatrix} \times \begin{pmatrix} \text{Planck's} \\ \text{constant} \end{pmatrix} = \begin{pmatrix} \text{Curviness of} \\ \text{psi wave} \end{pmatrix} + \begin{pmatrix} \text{Stored energy} \\ \text{of psi wave} \end{pmatrix}.$$

Without going into the mathematics, we can understand the meaning of each part of this equation. The first part, to the left of the equal sign, is closely related to Planck's relation, which as you recall relates the rate of ticking of a particle's internal clock to the particle's total energy. The 'curviness' part represents the energy of motion of the electron, as suggested by de Broglie. A greater energy of motion is associated with a greater momentum, and this is associated with shorter full-cycle length and thus a more curvy psi wave, as illustrated in FIGURE 6.3. The part of the equation at the far right accounts for the energy stored in the particle by virtue of its location, like the bob in a pendulum clock. Thus, Schrödinger's equation is a restatement, in quantum language, of the energy equation presented earlier; the total energy equals the sum of the energy of motion and the stored energy.

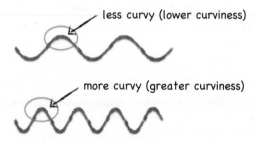

Figure 6.3 Illustrating curviness. The upper psi wave is more strongly curved at its crests and troughs than is the lower psi wave. The upper example has greater curviness—that is, it is more tightly bent.

How is the quantum wave related to probability?

Max Born, in 1926, first made the suggestion that the quantum possibility wave is tied closely to the probability of detecting the electron at a particular location. Born's Rule says the probability equals the *square* of the value of the psi wave at that location—that is, ψ^2. The electron is most likely to be detected in a region where the wave has a large. For a location where the wave has zero value, the probability of detecting the electron there is zero. Notice I did not say, "the probability for the electron to *be* at a certain location." That would imply the electron already had a definite position in space before it was detected. We know from our earlier discussion that such a statement is faulty thinking and would lead to incorrect predictions.

How is the psi wave related to the state arrow diagrams introduced in Chapter 4? Only two possible outcomes were considered in those examples, and we could draw a state arrow pointing to one or the other, or at some angle between them. The in-between state arrows represented quantum superposition states. We used Born's Rule, which says the probability of detecting a particular outcome is the square of the length of the possibility arrow that points to that outcome.

Now we must consider an infinite number of possible outcomes, because we are considering an electron that may be detected at any number of possible locations in space. But we can't easily draw a figure with an arrow pointing between infinitely many possible outcomes! In this case, the psi wave is comprised of the infinitely many quantum possibilities, one at each location in space. The psi wave is a representation of the fact that possibilities for traveling different paths can interfere and thereby change probabilities of measurement outcomes.

What is an example of Schrödinger's equation in action?

Imagine a classically described particle, such as a ball or a snowboarder, moving and then running up the side of a valley, as shown in FIGURE 6.4(i). Let's consider a situation in which there is no friction.

As the particle moves up the right side of the valley, it slows and comes to rest for an instant; all of its energy of motion has been converted to stored energy. As the particle turns around and begins to move back downhill, it regains energy of motion, then passes through the lowest point in the valley at maximum speed, starting the cycle again on the valley's other side.

If there is no friction, this oscillating motion continues indefinitely. If you were to close your eyes for a long while and then at some arbitrary time open them, what would be the probability of seeing the particle at any particular location? That probability will be greatest where the particle spends the most time—namely, near the turnaround points where the speed is momentarily zero. The diagram in FIGURE 6.4(iii) plots this probability, as predicted by classical physics. The probability is smallest at the center, where the particle's speed is greatest, so it spends the least time there. The probability is zero in the outer regions, where the particle does not have enough energy to go.

Now consider an *electron* that moves in a similar 'valley,' as in FIGURE 6.4(ii). The valley is made by placing negative

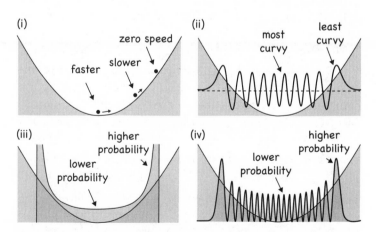

Figure 6.4 Classical (i, iii) and quantum (ii, iv) representation of motion of an object in an energy valley. View (ii) shows a stationary psi wave representing an electron, with the zero of the wave indicated by the dashed line. View (iv) shows that the square of the psi wave's value at each point indicates the probability of finding the electron at that location.

electric charge on two flat walls on opposite sides of the region where the electron travels. As it approaches either wall, the electron feels repulsive forces pushing it back toward the center. This creates an oscillating motion of the electron's position. The total energy of the electron is constant as it oscillates back and forth. Therefore, at a location where the energy of motion is zero, the stored energy is a maximum. This occurs at the classically predicted turnaround points. Therefore, Schrödinger's equation predicts that, in the regions of the turnaround points, where the electron has the smallest speed according to the classical picture, the psi wave will have a minimum of curviness. This can be seen at the location labeled 'least curvy.' The curviness of the psi wave is greatest at the center location, where in the classical picture the electron is moving with the greatest speed. 6.4

The crests and troughs of the psi wave in FIGURE 6.4(ii) do not move as time passes, unlike in the previous examples of de Broglie waves. This is because, as the electron oscillates right and left, it has two de Broglie waves associated with it: one moving right and one moving left. These two waves interfere constructively at some locations and destructively at others. This creates a nonmoving, stationary wave pattern.

The probability of finding the electron at any specific location, shown in FIGURE 6.4(iv), is specified by the square of the psi wave, according to Born's Rule. The probability is highest at the turnaround points at both extreme ends of its motion. This fact agrees with the classical prediction shown at the left in FIGURE 6.4(iii). The main difference between the classical and quantum predictions for the probabilities is the presence of interference in the latter case. Such interference occurs only if the electron's motion is a unitary process—that is, it is not being detected or leaving behind any permanent trace of its whereabouts. If it did leave such a trace—say, by reflecting a bright light shone onto it—then the interference would disappear and the probability pattern would become the same as the classical prediction.

How does a quantum particle get through locations of zero probability?

A remarkable aspect of the behavior of a quantum particle between two 'valley' walls is that there are locations along the valley where the probability of finding the particle is actually equal to zero. Those locations are the places in FIGURE 6.4(iv), where the curvy line touches the bottom edge of the figure. This means the particle can be found at various places despite the fact there are locations in between where the particle can never be found. There is no intuitive explanation for how this can be. This fact reinforces our realization that the position of a quantum object is not an attribute of the object that has any meaning in the absence of its being measured.

What is Heisenberg's Uncertainty Principle?

Many people with only a little familiarity with quantum physics have heard of **Heisenberg's Uncertainty Principle**. A common statement of the Uncertainty Principle is that "You can't measure precisely both the position and momentum of a particle at the same time." Although this statement is correct, we have to keep in mind that a quantum particle doesn't *have* a position or momentum (that is, velocity) before it is measured. Thus, to 'measure' either one is not to reveal a preexisting value; rather, the process of measuring one or the other elicits a particular result that gets 'decided' during the measurement process itself. So we need a more delicate statement of Heisenberg's Uncertainty Principle.

Schrödinger's equation provides the insight we need. Let's consider an electron moving freely with no forces acting on it. In this case the stored energy plays no role. All the electron's energy is energy of motion. Then Schrödinger's equation says the rate of change of the psi wave at each location is proportional to its curviness—that is, how tightly it is bent.

Consider an electron whose psi wave at a particular time is confined to a small region, as shown by the vertical dashed

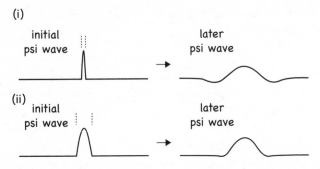

Figure 6.5 (i, ii) A more tightly confined psi wave (i) spreads out subsequently more rapidly than one that is less tightly confined (ii).

lines in FIGURE 6.5. The psi wave must equal zero at the edges of the confining region and everywhere outside this region. FIGURE 6.5(i) shows the predicted evolution of the psi wave if it is confined initially to a very small region. To fit into this small region and be zero at its edges, the psi wave must be strongly curved or bent; that is, it must have a large amount of curviness. According to Schrödinger's equation, this means the electron has a high possibility of a large amount of energy of motion; that is, it may have large momentum. Therefore, if we wait for a short time after the initial time and then measure the electron's position, we would not be surprised to find the particle located at a great distance from the starting location.

In contrast, if the psi wave is confined initially to a wider region, as in FIGURE 6.5(ii), Schrödinger's equation predicts that the wave spreads subsequently more slowly because it has lower curviness and thus less possibility of large momentum. Schrödinger's equation yields a precise statement of these facts:

> The smaller the region to which a psi wave is confined initially, the quicker it spreads out subsequently, giving the electron a higher possibility of being detected far from the starting location.

Note that in either case, the electron may still be found in or near the initial region; that is, it may or may not exhibit large momentum. There is a spread of possible momentum values that are implicit in the initial shape of the psi wave. The inherent spread of possible momentum values depends on how tightly confined the electron is initially. Putting these arguments together, we can restate the needed principle:

Heisenberg's Uncertainty Principle: The more precisely you can specify the position of a particle, the less precisely you can specify its momentum, and vice versa.

By specify, I mean 'specify the range of possible values of.' The range of possible values is called the *uncertainty*. This word implies something about what you can know or cannot know—that is, how uncertain you are about the outcome of a subsequent measurement. This kind of language might seem to imply you can, in principle, know more than you actually do, and the limitation we are talking about here is a result of only your ignorance or lack of information. This is not the case. The limitation embodied in the Uncertainty Principle is a fundamental physical fact about quantum particles, not a reflection of a person's inability to make precise measurements.

In fact, Heisenberg originally used the German word 'ungenauigkeit,' which means inexactness or vagueness, rather than 'uncertainty,' to describe his principle. This is closer to the true meaning of Heisenberg's Principle than is the word 'uncertainty.'

The discussion in this section shows that the Uncertainty Principle is not an added feature or postulate of quantum theory that might seem to come out of nowhere. Rather, it is a direct consequence of the nature of waves, if we take seriously de Broglie's idea that a mathematical quantum possibility wave should be associated with each particle.

Is it correct to say an electron is both a particle and a wave?

No. It's not correct to say an electron is both a particle and a wave. It's not even correct to say that sometimes it behaves like a particle and sometimes like a wave. It is best to say the pattern of possibilities that Schrödinger called psi, and the one associated with the electron, is a mathematical wave. And this wave determines the probability of observing the electron at certain locations and with certain velocities.

If psi can be analyzed as a wave, then what became of the particlelike nature of the electron? This exhibits itself only when you attempt to measure the particle's location or velocity. For example, if you place something like a piece of photographic film in the vicinity of the electron, and if the electron strikes the film's surface, it will leave a small permanent spot on the film as a result of the electron's energy being deposited there. It will leave only one spot, because there is only one electron. So, although we use the word 'particle' when referring to electrons, we really mean 'quantum particle,' which is nothing like the classical concept of a particle.

Notes

1   The tick marks on an electron's quantum ruler move at one-half the speed the electron moves.
2   Readers who already know some physics might recognize this equation as Newton's Second Law ($F = m \times a$)—that is, force equals mass times acceleration.

7

MILESTONES AND A FORK
IN THE ROAD

What aspects of quantum physics have we seen so far, and what topics should we discuss next?

We find ourselves at a fork in the road on the way to understanding quantum physics. In a historical progression, it would make sense next to discuss how Schrödinger's equation describes the properties and behaviors of atoms, as was Schrödinger's original motivation. This area of study is called *quantum mechanics*. This topic is of practical importance because much of modern technology, including computer engineering and material science, is based on understanding how atoms combine to create materials with special properties such as electrical conductivity, beneficial chemical properties, and useful mechanical properties.

On the other hand, some of the most recent and intriguing applications of quantum physics are in a new area called *quantum information science*. We encountered one example already in Chapter 3: quantum key distribution. Another great potential application of quantum information science is quantum computing, which is the art of building computers that exploit quantum-state superposition to carry out computations that cannot be performed nearly as efficiently otherwise. Quantum information science is also contributing greatly to basic research in areas as diverse as black-hole physics and thermodynamics.

The famous baseball player Yogi Berra said, "When you come to a fork in the road, take it." So I first take the road to quantum information and save further discussion of quantum mechanics for Chapters 11 and 12. Before diving into quantum information, it is useful to take stock of what has been discussed up to this point.

What milestones have we passed so far?

The ten milestones we have passed so far have two aspects: (1) the rules telling us how to calculate probabilities for outcomes of measurements and (2) the conceptual shifts we have been forced to make in our thinking about how the world behaves and how we should describe it.

Milestone 1: Inherent Randomness

The first milestone is the observation that, although classical physics rests on the assumption that experimental results are inherently reproducible, quantum physics recognizes that an experimental procedure, if reproduced and repeated exactly, may give two different results. This means Nature is not deterministic; it contains inherent randomness that cannot be eliminated by gaining more information about the situation at hand. For this reason, the concept of probability becomes an essential feature of the description of Nature.

Milestone 2: Measurement

The second milestone is that the idea of measurement has a meaning different than was supposed in classical physics. Rather than revealing the values of preexisting properties, as in classical physics, a quantum measurement *creates* or *elicits* an outcome, which very much depends on the particular measurement scheme used. Certain measurements are 'complementary,' in that performing one precludes the possibility of also performing the other.

Milestone 3: Quantum State

The next milestone is the realization that the way of describing the condition or state of a physical object needs to be modified from that used in classical physics. In classical physics, a state is a direct description of an object's properties, such as position, velocity, energy, or, for a light wave, polarization. In classical physics, there is a one-to-one correspondence between the state and outcomes of a measurement. In contrast, a quantum state is not in one-to-one correspondence with outcomes of measurement. Rather, the quantum state is the information needed to predict probabilities for outcomes of any conceivable measurement. No more detailed or precise specification can be given for a quantum object.

A quantum state describing a single object is a rather private affair; you can't make a copy of the state—a principle of quantum physics called the 'no cloning principle'—without destroying the original copy. And you can't determine the state by making measurements on that single object.

Milestone 4: Max Born's Probability Rule

Born's Rule tells us how to calculate probabilities for measurement outcomes from knowing a quantum state. For a quantum object that has only two possible outcomes for any given measurement, FIGURE 7.1 summarizes the geometry and terminology used for implementing the rule. The state Ψ and the two possible outcomes A and B that may occur are represented by arrows with lengths equaling one. The component arrows that combine to make the state arrow are called 'possibility arrows' and are labeled aA and bB. The lengths of the possibility arrows are numbers, a and b, which are called 'possibilities.' Their values, when squared, give the probabilities, a^2 and b^2, for observing each of the outcomes upon measurement.

When a state is recognized as existing partway between two possible outcomes, we say it is a 'superposition state.' There is no counterpart for such a state in a classical conception of Nature.

Milestone 5: Unity of Measurements and States

This milestone unifies milestones 2 through 4 in that it recognizes a deep and subtle connection between measurements in

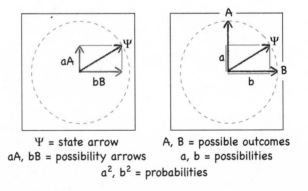

Ψ = state arrow   A, B = possible outcomes
aA, bB = possibility arrows a, b = possibilities
a^2, b^2 = probabilities

Figure 7.1 Elements of quantum theory: state arrows, possibility arrows, possible outcomes, possibilities, and probabilities.

one scheme and measurements in another scheme, and that the connection is made by the nature of the quantum state. For example, consider the polarization state of a single photon. The state arrow can be represented as the sum of two possibility arrows in the measurement scheme corresponding to the horizontal (H) and vertical (V) directions. Knowing that state arrow, you can deduce the possibility arrows in any other measurement scheme—for example, the scheme defined by the diagonal (D) and antidiagonal (A) directions. Thus, the concept of quantum state is more fundamental than a simple listing of outcome probabilities in a particular measurement scheme.

By repeating measurements many times for a limited number of different schemes, the quantum state can be determined experimentally. Because the quantum state can be inferred only indirectly from such a large set of diverse measurements, such a method for determining it is called quantum-state tomography.[1]

Milestone 6: Unitary Processes

When a quantum object undergoes a physical process during which no measurement is made and no permanent traces of the object's properties are created, the process is described as 'unitary.' During such a process, the changing of the state is represented by a reorientation of the state arrow relative to the possible outcome arrows.

An example is the polarization of a photon of light. In ordinary materials such as air, water, or glass, the polarization direction remains constant as the light travels. But, when traveling in certain substances, such as sugar water, the sugar molecules interact with the light and cause its polarization to change direction—that is, to rotate. This is described by drawing the state arrow with a new orientation while keeping the H-pol and V-pol outcomes directions fixed. This change of direction of the state arrow affects the probabilities for outcomes if the polarization is measured.

Another example of a unitary process is provided by an electron traveling toward a detector when there are two possible paths that lead to it. The electron being registered by this detector is a possible outcome. The two paths are represented by possibility arrows. If the two possibility arrows combine or 'interfere' in the right way, they will create a state arrow pointing to the detector outcome; then it is certain that the electron will be detected at that detector. But, if the two paths have slightly different lengths than considered previously, the same two possibility arrows may interfere in the opposite way and yield zero probability of the electron being detected at that detector.

Milestone 7: Planck's Energy–Time Relation

Planck's relation states that every quantum particle has associated with it a repeating oscillation in time, which I called an internal quantum clock. The time between ticks of this fictitious clock is the 'full-cycle time,' which is given by Planck's constant divided by the particle's energy.

For photons, Planck's relation states that, its energy content, E, is related directly to its frequency, f, which is an indication of the color of light. A photon's energy is given by Planck's constant multiplied by the frequency (that is, $E = hf$, where h is Planck's constant).

Milestone 8: de Broglie's Momentum–Length Relation

de Broglie's relation states that every quantum particle has associated with it a repeating length scale, which I called a quantum ruler. The distance between major tick marks on this fictitious ruler is the 'full-cycle length,' which is given by Planck's constant divided by the particle's momentum.

Milestone 9: Schrödinger's Equation for Possibility Waves

Combining milestones 7 and 8 for an electron allowed Schrödinger to define an equation that describes a wave of quantum possibilities that moves through space in an oscillating manner. The equation accounts for both energy of motion and stored energy of the electron, and its solution can represent a wide variety of electronic and atomic processes. The psi wave, symbolized as ψ, represents an infinite number of possibilities—one associated with every point in space. Born's Rule says that the probability equals the *square* of the value of the psi wave at that location—that is, ψ^2.

Milestone 10: Heisenberg's Uncertainty Principle

According to Schrödinger's equation, the smaller the region to which a psi wave is confined initially, the quicker it spreads out subsequently, giving the electron greater possibilities for being detected far from the starting location. This implies that the more precisely you can specify the position of a particle, the less precisely you can specify its momentum (that is, its velocity) and vice versa.

Note

1   My research group at the University of Oregon was the first to determine completely a quantum state of light, in 1993, and we introduced the term 'quantum tomography' into the physics lexicon.

8

BELL-TESTS AND THE END
OF LOCAL REALISM

Can experiments probe the nature of reality?

The year 2015 was a very good year for quantum theory, yet it was not such a good year for 'classical reality.' Since the inception of quantum theory in the 1920s, its implications regarding what can be thought of as 'real' have been a contentious issue. Although such an issue might seem to be in the realm of philosophy, it can be put to the test experimentally. Although there were famous debates between Albert Einstein and Niels Bohr addressing this question, the person who did the most to move it into the realm of experimental tests was John Bell. The types of experiments he proposed during the 1960s are now called Bell-test experiments, or simply Bell-tests.

A commonsense worldview is Local Realism. By 'worldview' scientists mean a grand, overarching conception about what the world is and how it works. *Local Realism* is the worldview that holds that physical objects carry with them definite properties or 'instructions' for how to respond when a measurement is performed on them, and that physical influences cannot travel faster than the speed of light. Bell proved theoretically that quantum theory is inconsistent with Local Realism. Bell also proposed ingenious experiments that could test directly whether the Local Realism worldview is tenable. During the decades since then, starting with the studies of John Clauser in 1972, many experiments of the Bell-test type have been carried out, with results always in agreement with quantum theory's predictions. Yet, until 2015, no experiment had been performed that didn't suffer from one or more technicalities, which a skeptic could invoke to argue the experiments were not definitive. In 2015, three different laboratories—two in Europe and one in the United States—performed experiments in which these technicalities were 'cured,' leaving virtually no doubt that Local Realism as a worldview is not correct.

This chapter explains the assumptions behind Local Realism, and how the Bell-tests invalidate it. The invalidation of such an ingrained, commonsense view of reality is a highly curious and even shocking result. It brings home the truly revolutionary nature of quantum physics. The overthrow of Local Realism also points the way to new quantum technologies, such as quantum computers. The Bell-tests rely on measurements of correlations, so let's first explore this concept.

What is correlation and what does it tell us?

Correlation is the extent to which two things (properties, behaviors, events, and so on) are related to each other. For example, if the stock prices of two corporations tend to rise and fall together, we say they are correlated. If they tend to rise and fall oppositely, then we say they are correlated negatively. If they tend to change in completely unrelated ways, we say they are uncorrelated.

There can be different reasons a correlation exists. In the stock example, it may be that both corporations are affected similarly by some external factor such as the cost of oil (a common cause), it may be that one or the other company's situation affects directly the situation of the other (a causative link), or it may be that the observed correlation arises by coincidence and is purely accidental (no common cause or causative link).

In science, correlation between two physical properties or behaviors is usually noticed by a mathematical analysis of two sets of numerical data obtained by observations or measurements. The observation of a correlation is often the starting point for further investigation. If a common cause or a causative link can be found, then we have learned something. But keep in mind that the mere observation of a correlation doesn't tell you much by itself. You need to look deeper for a cause, if there is one.

Physicists have discovered that correlations we can observe for 'quantum' objects can be of a very different kind than any correlations we can observe for 'classical' objects. Measuring correlations between two separate but related quantum objects raises deep questions about the nature of reality.

What is an example of correlated properties?

John Bell, the most important innovator in the study of quantum correlations, gave a fun example involving a professor colleague of his:

> Dr. Bertlmann likes to wear two socks of different colours. Which colour he will have on a given foot on a given day is quite unpredictable. But when you see [FIGURE 8.1] that the first sock is pink you can be already sure that the second sock will not be pink. Observation of the first, and experience of Bertlmann, gives immediate information about the second. There is no accounting for tastes, but apart from that there is no mystery here.[1]

Figure 8.1 Bertlmann's socks (drawing by John Bell, used with kind permission of *The European Physical Journal* [EPJ]).

In this case we could reasonably presume a common cause for a particular property—color—of Bertlmann's socks. For example, perhaps the laundry where he sends his socks for cleaning has a washing machine that operates only if it contains, no more than one sock of any color. In any case, as Bell says there is no real mystery here—just a simple correlation of sock properties, which likely has roots in a common cause.

What is an example of correlated behaviors?

Correlation may occur for behaviors as well as for properties. As an example, two ballet dancers are scheduled to perform solo dances in two separate halls beginning at 8 pm on the same day. They meet in advance to develop a complex series of dance moves, and they agree that both will execute the same moves in their separate performances. When they later perform their dances, each following the same 'program' or 'instruction set,' it may look to each separate audience like a random, improvised dance. Say that an observer named Alice was in one audience and Bob was in the other. If Alice and Bob were later to compare their observations of the dances, they would conclude there was a strong correlation between them. They would conclude that most likely the two dancers had agreed in advance to perform identical memorized dances. This is an example of a preexisting common cause.

Or the observers might look for alternative explanations other than the dancers being preprogrammed. They might suspect the two dancers wore hidden earphones and microphones that allowed them to communicate during their dances. If that were the case, the dancers would not need to follow a memorized program; they could agree quickly in real time which step or move to make next, then both execute it. In this case, the dances would be truly random (on the assumption that people do have free-will!), yet still perfectly correlated. This is an example of a common cause that did not preexist but was created at random and communicated instantly.

How can correlations be quantified?

Rigorous analysis of correlations is based on numbers. To represent the outcomes of correlation experiments, we assign specific numbers to specific observations or measurements. Let's say that Alice and Bob took one photograph every second of their dancer, then they each made an ordered list that recorded for each photo whether their dancer's left arm was up (denoted +1) or down (denoted –1), and whether their dancer's left leg was up (denoted +1) or down (denoted –1). In TABLE 8.1, each row is for a single photograph of each dancer. Alice's list of observed arm and leg positions might look like that in the first and second columns, named AA for 'Alice sees arm' and AL for 'Alice sees leg.' Bob's list might look like that in the third and fourth columns, named BA for 'Bob sees arm' and BL for 'Bob sees leg.' I refer to AA, AL, BA, and BL as 'outcomes.' (We won't consider the movements of the right arms and legs.)

Table 8.1 Lists of Observations of Two Dancers' Arm Positions (AA and BA) or Leg Positions (AL or BL)

Alice sees arm AA	Alice sees leg AL	Bob sees arm BA	Bob sees leg BL	AA×BA	AA×BL	AL×BA	AL×BL
1	−1	1	−1	1	−1	−1	1
−1	−1	−1	−1	1	1	1	1
1	1	1	1	1	1	1	1
−1	−1	−1	−1	1	1	1	1
1	−1	1	−1	1	−1	−1	1
−1	1	−1	1	1	−1	−1	1
−1	−1	−1	−1	1	1	1	1
−1	−1	−1	−1	1	1	1	1
−1	−1	−1	−1	1	1	1	1
1	1	1	1	1	1	1	1
−1	−1	−1	−1	1	1	1	1
1	1	1	1	1	1	1	1
−1	1	−1	1	1	−1	−1	1
1	−1	1	−1	1	−1	−1	1
−1	1	−1	1	1	−1	−1	1
1	1	1	1	1	1	1	1
1	1	1	1	1	1	1	1
−1	1	−1	1	1	−1	−1	1
1	−1	1	−1	1	−1	−1	1
−1	−1	−1	−1	1	1	1	1
1	−1	1	−1	1	−1	−1	1
−1	1	−1	1	1	−1	−1	1
−1	1	−1	1	1	−1	−1	1
−1	1	−1	1	1	−1	−1	1
−1	−1	−1	−1	1	1	1	1
		Correlation =	1	0.04	0.04	1

The bottom row gives the correlation of each pair.

To quantify the amount of correlation between any two types of observations—that is, any two sets of outcomes—we can do the following: multiply each number in Alice's list by the corresponding number in Bob's list. If the two lists are perfectly correlated—that is, identical—then for each location where +1 occurs in Alice's list, +1 will also appear in Bob's. The same is true for all −1 entries in the two lists. Therefore, the products would always yield +1, because (+1) × (+1) = +1 and (−1) × (−1) = +1. Such a case is illustrated in the fifth column labeled AA × BA, which shows the product for each observation of arm by Alice and arm by Bob. At the bottom of the AA × BA column is the average of all the product numbers above it. This average, also called the correlation, equals +1. I constructed the 'arm' data to illustrate positive correlation between AA and BA; as you can see, the entries in these lists are identical. This is consistent with the two dancers performing identically.

On the other hand, if two outcomes are uncorrelated, then each +1 entry in Alice's list might be multiplied by a +1 or a −1 from Bob's list. So the products may be +1 or −1 in this case. For uncorrelated outcomes, these possibilities would occur with equal likelihood; therefore, the values of these products will tend to average to zero. Such a case is illustrated in the sixth column in TABLE 8.1, labeled AA × BL, which shows the products for each observation of arm by Alice and leg by Bob. At the bottom of this column is the average of all the product numbers above it, which equals 0.04, which is close to zero. I constructed the data specifically to illustrate near-zero correlation between the 'arm' and 'leg' outcomes AA and BL. This means that for the dance they performed, the (left) arm movements were uncorrelated with the (left) leg movements.

The seventh column in TABLE 8.1 shows that AL and BA are also nearly uncorrelated. The eighth column shows that, in this example, AL and BL are perfectly correlated, also as expected, because they dance identically.

Now, to continue our story, let's say that a month later Alice and Bob again attend parallel performances by two dancers at two different venues. The dancers are new, and Alice and Bob have no idea what dances will be performed. They again record the movements of the separate dancer's left arm and leg, and afterward create the data lists in TABLE 8.2. They see, as before, the arm movements are uncorrelated with the leg movements; but they notice a new behavior. The correlation of AA and BA for these performances has a value of −1. That is, the two dancers' arm movements are perfectly negatively correlated; each time one dancer raises the (left) arm, the

Table 8.2 Lists of Observations and Correlations of Two Dancers' Arm (AA and BA) or Leg (AL or BL) Positions

Alice sees arm AA	Alice sees leg AL	Bob sees arm BA	Bob sees leg BL	AA×BA	AA×BL	AL×BA	AL×BL
1	1	−1	−1	−1	−1	−1	−1
−1	−1	1	1	−1	−1	−1	−1
1	−1	−1	1	−1	1	1	−1
1	−1	−1	1	−1	1	1	−1
1	1	−1	−1	−1	−1	−1	−1
1	1	−1	−1	−1	−1	−1	−1
1	−1	−1	1	−1	1	1	−1
1	−1	−1	1	−1	1	1	−1
−1	−1	1	1	−1	−1	−1	−1
1	1	−1	−1	−1	−1	−1	−1
1	1	−1	−1	−1	−1	−1	−1
−1	1	1	−1	−1	1	1	−1
−1	−1	1	1	−1	−1	−1	−1
−1	1	1	−1	−1	1	1	−1
1	−1	−1	1	−1	1	1	−1
1	1	−1	−1	−1	−1	−1	−1
1	−1	−1	1	−1	1	1	−1
1	−1	−1	1	−1	1	1	−1
−1	−1	1	1	−1	−1	−1	−1
−1	−1	1	1	−1	−1	−1	−1
1	−1	−1	1	−1	1	1	−1
−1	−1	1	1	−1	−1	−1	−1
−1	1	1	−1	−1	1	1	−1
−1	−1	1	1	−1	−1	−1	−1
−1	1	1	−1	−1	1	1	−1
	Correlation =	−1		−0.04	−0.04		−1

other lowers the (left) arm. The same perfect negative correlation is also observed for leg movements of the two dancers. Whenever Alice sees AL = +1, Bob sees BL = −1, and vice versa, giving a correlation of −1.

To summarize, we define the correlation of two lists as the average of the products of the corresponding list entries. If the correlation calculated this way is positive, we say there is (positive) correlation; if the calculated correlation is zero, we say there is no, or zero, correlation (the lists are uncorrelated); if the calculated correlation equals a negative number, we say there is negative correlation. Keep in mind that calculated correlations are reliable only if a large enough number of test cases are observed.

What is the difference between classical correlation and quantum correlation?

In the case of socks, we can talk about their classical properties—namely, their color—properties that existed before we observed them. In the case of dancers, we can talk about programs or sets of instructions, which may have been predetermined or may have been created randomly on the spot and communicated quickly. We might not know precisely the cause (if any) behind the correlations of the sock colors or dance moves, but we have no trouble making up plausible scenarios that could explain the correlations we observe.

When dealing with quantum objects such as photons or electrons, explanations of correlations are not so simple. Experiments have been carried out for which the observed correlations have *no* intuitive, commonsense explanation that can be given in terms of preexisting properties or instruction sets.

This point deserves emphasis. Recall in earlier chapters I argued that photons and electrons behave as if they don't have definite, preexisting properties, such as polarization or position. I tried to convince you that this conclusion was inescapable given the evidence I presented. But scientists (and you!) should always be skeptical when presented with such a radical idea that goes against common sense. ("What do you mean, an electron has no position until you observe it? Ridiculous!") You should ask for proof.

The experiments that I describe next offer what many physicists take as strong experimental evidence that, in at least some experimental situations, quantum objects cannot be said to have preexisting properties or instruction sets. If this conclusion is valid, then it means that, in general, quantum objects do *not* have preexisting properties or instruction sets. Some people consider this to be the most profound of all scientific discoveries. If true, it certainly is revolutionary!

As I will show, the proof of this claim involves only experimental evidence combined with simple data analysis and commonsense logic. The ideas and theory of quantum mechanics are not needed at all for this proof. However, I need to clarify carefully what is meant by preexisting properties, instruction sets, and realism in this context.

What is realism and how can we test it experimentally?

The idea of 'realism' is a key concept in the philosophy of science and it can be defined in various ways. When I speak of realism, I mean a specific concept that lay at the foundations of physics theories during the classical pre-1900 era. A definition is as follows:

> *Realism* is the idea or worldview that physical entities exist and have definite properties and/or instructions for behaviors, regardless of whether they are observed, and independent of anyone's beliefs or theories.

The word 'realism' represents a belief that objects *really* do have concrete, preexisting properties and/or instructions that are inherent or innate to themselves. The instructions could be, for example, the laws of classical physics, such as Newton's force and acceleration equation.

When adopting the worldview of realism, we envision that correlated objects behave as they do because they have preexisting properties or they share instruction sets. And we assume that measurements that we did *not* carry out would certainly yield *some* outcome, independent of other measurements we did or did not perform. This viewpoint fits well with our commonsense understanding of things such as socks and dance performances. But, as we shall see, it does not work as a description of how quantum objects behave.

How can we carry out a rigorous test of whether realism is a valid concept? First, note that if realism is to be thought of as valid, then it must *always* be valid. That is, it must provide a valid model for all experiments that are carried out. The idea that John Bell had in 1964 is as follows: Let's try to think of a mathematical relation involving measured outcomes that must *always* hold true if realism is valid. Such a relation is called a 'Bell Relation.' The idea is that if we can find such a relation, and then if we find even a single experiment that gives results in disagreement with the Bell Relation, then we will have shown that realism is suspect as a model of nature.

Setting the stage for experimental tests of realism

Before discussing the correlation experiments that were carried out using photons, I should point out some important features of the correlations that can occur for 'classical' events such as dance performances. It might be clear already that the correlation between any of the four outcomes AA, AL, BL, and BL cannot exceed +1, nor can it be less than −1. This is true simply because these four outcomes themselves always have values +1 or −1, and the correlation equals the average of their products, as we discussed for TABLE 8.1 and 8.2. A good way to visualize this statement is shown in TABLE 8.3. The first four columns list every possible combination of the two dancers' arm positions (AA and BA) and leg positions (AL and BL).

For every combination, the products are also shown, and they all equal +1 or −1.

There are several interesting relations we can verify by looking at all the possible combinations in TABLE 8.3. For example, consider the sum of the four products, which I denote by S, expressed as S = AA × BA + AA × BL + AL × BA + AL × BL. You can see in the table that this sum cannot exceed +4. We can summarize this statement as the relation S ≤ 4, which reads, "S is less than or equal to 4." This kind of relation is called an inequality.

A different quantity that we can consider is one I will call 'the curious quantity,' or Q. It is defined by Q = AA × BA + AA × BL + AL × BA − AL × BL. That is, to calculate Q, we add

Table 8.3 All Sixteen Possible Combinations of Two Dancers' Arm (AA and BA) and Leg (AL and BL) Positions

AA	AL	BA	BL	AA×BA	AA×BL	AL×BA	AL×BL	S	Q
1	1	1	1	1	1	1	1	4	2
1	1	1	−1	1	−1	1	−1	0	2
1	1	−1	1	−1	1	−1	1	0	−2
1	1	−1	−1	−1	−1	−1	−1	−4	−2
1	−1	1	1	1	1	−1	−1	0	2
1	−1	1	−1	1	−1	−1	1	0	−2
1	−1	−1	1	−1	1	1	−1	0	2
1	−1	−1	−1	−1	−1	1	1	0	−2
−1	1	1	1	−1	−1	1	1	0	−2
−1	1	1	−1	−1	1	1	−1	0	2
−1	1	−1	1	1	−1	−1	1	0	−2
−1	1	−1	−1	1	1	−1	−1	0	2
−1	−1	1	1	−1	−1	−1	−1	−4	−2
−1	−1	1	−1	−1	1	−1	1	0	−2
−1	−1	−1	1	1	−1	1	−1	0	2
−1	−1	−1	−1	1	1	1	1	4	2
							Average	≤ 4	≤ 2

For every combination, the products are also shown, as are the sum S = AA × BA + AA × BL + AL × BA + AL × BL; and the quantity Q = AA × BA + AA × BL + AL × BA − AL × BL.

the first three products and subtract the fourth product. For example, in the first row, the products are 1, 1, 1, 1. Therefore, the curious quantity Q equals $1 + 1 + 1 - 1 = 2$. We can check that, in every case, the curious quantity equals +2 or –2, so its average is always less than or equal to 2. Therefore, we can confirm the inequality relation $Q \leq 2$.

I don't claim it should be obvious at this point why these inequalities involving S and Q are of interest for experimental tests of realism. This is why I refer to Q as the curious quantity. It turns out, as I show next, that the inequality $Q \leq 2$ leads directly to a Bell Relation. Then, if we find even one experiment that gives results in violation of the Bell Relation, then we will have shown that realism as a model of nature is suspect.

The needed Bell Relation is deduced by noting that, in any experiment, there are only sixteen combinations of observation results; these are listed in TABLE 8.3. Then, note that no matter which mixture of these sixteen results actually occurs, if all four outcomes are measured repeatedly and averaged, the average value of the curious quantity Q cannot exceed +2. This fact is indicated at the bottom of the far-right column. This conclusion is crucial, so I will restate it here:

> If all four outcomes (AA, BA, AL, BL) are measured repeatedly, the average value of Q cannot exceed 2. That is,
> $$\text{Ave}(Q) \leq 2.$$

Now, according to the way averages work, we can obtain the average value of Q by finding the average of each of the four products separately, then adding or subtracting them according to the original meaning of Q. That is,

$$\text{Ave}(Q) = \text{Ave}(AA \times BA) + \text{Ave}(AA \times BL) + \text{Ave}(AL \times BA) - \text{Ave}(AL \times BL).$$

Here, Ave( ) means "average of whatever is within the parentheses."

Notice that the averages in this equation are nothing other than the correlations of the four outcomes presented in TABLES 8.1 and 8.2. So we can determine the value of Ave(Q) by calculating the four correlations from the observed data and then combining them, as in the previous equation. The main message is that Ave(Q) is a particular way to characterize the amount of correlation among four outcomes.

It is crucial to realize that the relation Ave(Q) ≤ 2 holds true not only for dance performances, but for *any* correlated events that can be represented using completely filled-in data lists of the type in TABLES 8.1 and 8.2. That is, if we represent any complete set of measurements using our +1/−1 scheme (which we can always do), then Ave(Q) ≤ 2 will always be true. This statement doesn't depend on the worldview one has; it is purely a consequence of mathematics.

What if we can make only partial measurements?

How does our reasoning about correlations change if, for some reason, Alice and Bob are not able to make complete observations such as those in the fully filled-out tables in TABLES 8.1 and 8.2? That is, what if they can each observe only one of the outcomes during each measurement?

This limitation on measurement ability becomes important when measuring quantum objects, because for such objects you cannot measure all outcomes at the same time. In Chapter 2 I explained that Niels Bohr used the term *measurement complementarity* for this fact. For example, the particular way in which you set up a calcite crystal to measure a photon's polarization defines a particular *measurement scheme*. The V-pol or H-pol observed outcome is complementary to the D-pol or A-pol observed outcome. Once you have made the first measurement, you can't change your mind and go back to make a different measurement to learn something different. This is because a measurement not only gives you information, it also changes the quantum state of the object being measured. So the choice you make in the present affects what can be learned in the future.

Before describing the situation with quantum measurements, let's consider a 'classical' scenario in which partial measurements are made. Consider again the two dancers in separate theaters, with Alice and Bob in one or the other. The dancers want to spice up their performances, so they add a special effect. The stage is dark, so the audience can't see the dancers except when a strobe light flashes momentarily, freezing the dancer's motion in that instant. For each dancer there is one strobe light, which can be aimed either at the left arm or the left leg of the dancer. Alice and Bob each have in front of them two buttons: one labeled 'arm strobe' and the other, 'leg strobe.' They agree to push one or the other button at the same instant—say, once each second. Each can choose independently at random, with no preference, whether to illuminate the arm or the leg. Roughly one-half of the time they will both push a like-labeled button (arm–arm or leg–leg).

This means Alice and Bob can each observe only an arm or leg position at a given time. They record arm up (+1) or down (–1) or leg up (+1) or down (–1) each time a strobe flashes. Their recorded data might look like those shown in TABLE 8.2, but with each row containing only one entry from Alice and one from Bob, instead of two for each as before. And each row will contain only one product rather than four.

The good news is that having only part of the total information does not prevent them from calculating the correlations, which are given at the bottom of each product column. For example, if there are four hundred rows, then about one hundred of them will contain a value for AA × BA, so a meaningful average value for this product can be obtained. The same holds for the other three products.

Then, by adding the first three averages of products and subtracting the fourth average at the bottom of the figure, the value of Ave(Q) can be obtained even when each outcome is measured only part of the time. The value of Ave(Q) obtained by this procedure will be the same as that obtained if all outcomes were measured every time. At least, this statement is true for classical observations such as dancers' movements, because we know that nothing out of the ordinary is occurring for the arm and leg not being observed at a given time; each is either up or down. As John Bell said, "There is no mystery here."

So we can conclude: If realism is a valid worldview (meaning the unobserved arm and leg movements behave the same regardless of whether they are being observed), and if there is no communication between the two dancers and the persons or mechanisms operating the strobes (which could conceivably allow them to conspire and chose their movements just before each strobe is flashed), and if Alice and Bob are not being controlled secretly by some mysterious outside agent, then the average of the curious quantity Q determined by these observations cannot possibly exceed 2. That is, again, Ave(Q) ≤ 2. This inequality is the Bell Relation that we need.

The difference between the previous case and this new situation is that in the previous case, no assumptions were needed—just a mathematical proof—because everything was measured all the time. For the new conclusion just made, the fact that only one measurement is being made at a given time by both Bob and Alice requires us to make additional assumptions, which we discuss next.

What prevents communication between the two sides of the experiment?

In previous discussions, I implied it was legitimate to assume there was no communication between the two dancers and the persons or mechanisms operating the strobes. But how can we be certain of that? Scientists need to be on the lookout for physical mechanisms that might be at work, even if the mechanisms might operate by physical processes as yet unknown. (For example, a scientist of the sixteenth century would be mightily surprised to learn that two people could communicate nearly instantaneously across a distance of, say, a thousand miles. Now we just whip out our mobile phones.) There is a way we can set up our Bell-test experiment and be reasonably certain there was no communication: by appealing to the principle of 'causality' as it is now understood in the context of Einstein's theory of relativity—a theory that is on very firm footing.

The basic notion of *causality* is simple enough. Some physical influence causes some physical effect. For example, when lighting strikes, it makes the sound of thunder. Sound, which is a disturbance of ambient air pressure, travels at a speed of 344 meters per second. So a person 344 meters from the lighting strike hears thunder one second after the strike, whereas a person 688 meters from the lighting strike hears thunder two seconds after the strike. You can think of an imaginary 'sphere of influence' expanding at the speed of sound from the location of the lightning strike. At any particular time, only those people and objects within the sphere of influence can be affected by the sound caused by the strike. An event such as a lightning strike is a 'cause,' in that it may cause other events (such as your eardrums vibrating) at distant locations sometime in the future.

Einstein's great insight was that, in Nature, there exists a fastest possible kind of disturbance that can carry causes from one place to another, and this fastest possible disturbance is light. The speed of light equals about 3×10^8 meters per second. This is the absolute 'speed limit' for all entities and influences in the universe, as far as we know. For example, the light generated at the strike point travels to your eyes at the speed of light, and no disturbance can possibly travel any faster. Therefore, in the split second after the strike and before the light reaches your eyes, from your perspective it is truly as if the strike never occurred. As the light travels from the strike point, a 'causal sphere of influence' expands at the speed of light from this point. At any particular time, only those people within this expanding sphere can know about the strike; those outside the sphere cannot (yet) be affected in any way by the existence of the strike. That is, the strike cannot be a cause of any event that may occur outside the expanding causal sphere of influence. We can state a principle of relativity theory:

> **Principle of Local Causality**: An event or cause that occurs at a particular time can have no physical effect on any object that is currently outside of the causal sphere of influence, which expands from the event at the speed of light.

This principle, which is a foundation of Einstein's relativity theory, has been tested countless times, and has never been seen violated in any experiment. Even the force of gravity cannot travel faster than light.[2]

The speed of communication is limited by the Principle of Local Causality. Communication occurs through cause and effect. No information can be transferred from point A to point B unless some physical cause at point A creates a physical effect that is felt at point B. A time lag always exists between a message being sent and its reception, and the minimum lag is set by the speed of light.

This fact gives us an iron-clad way to make sure that Bell-test experiments are carried out while knowing there is no possibility of communication between the objects being measured or between the persons or mechanisms determining which outcome to measure for each object. We should place them very far apart and have the two experimenters, Alice and Bob, perform their measurements within a short time interval during which no signals of any kind can be sent and received by them.

What is Local Realism?

It's convenient to put a name to the worldview I have been describing thus far. It is called 'Local Realism,' and it is the combination of two ideas we discussed previously. *Local Realism* is the worldview that combines Realism and the Principle of Local Causality. In this assumed worldview, entities have definite, preexisting properties or behavioral instructions regardless of whether they are observed, and events can have no physical effect on any other event that is currently outside of the first event's causal sphere of influence, which expands at the speed of light.

What kinds of experiments can put an end to Local Realism?

What experimental results might have convinced Einstein that he needed to give up his local, realistic worldview? It turns out that by measuring the polarization of many pairs of photons that are correlated, physicists have obtained results that almost certainly prove that the idea of Local Realism is untenable, as I now explain.

Following the pioneering experiments by John Clauser and other researchers, in 1982, Alain Aspect, Jean Dalibard, and Gérard Roger carried out an experiment in Paris that is recognized as a milestone in testing the Bell Relation. In their experiment, these researchers observed pairs of photons that were created when an electron in a higher energy state of a calcium atom loses energy and drops to a lower energy state. If the electron does so in two steps, or 'quantum jumps,' then two photons are created, which travel away from the atom in different directions.

Because energy and momentum are constant before and after the photons are emitted, the two photons have correlated observable properties. For example, if one photon is observed to have a particular polarization, then the other will be observed to have a different particular polarization. Quantum theory predicts perfectly the observed correlations of polarization.

Instead of relying on quantum theory to understand the polarization correlations, let's pretend we are experimentalists working before 1900—before quantum theory was known. You can be Alice and I will be Bob. Because we are clever physicists, we have found a way to carry out an experiment as shown in FIGURE 8.2. We each have one light detector and we share a source of light. (I won't say photons because we are pretending we don't know quantum physics.) The light source is a group of atoms contained in some region. We shine light onto the atoms in the source to impart energy to them. We see that red light is emitted by the source toward our detectors, which register detection events, which we call 'clicks' (similar to how a Geiger counter sounds when it detects a radioactive particle). We notice that both detectors seem to click at random, but that they always click at the same time. From this, we surmise the source is emitting light in a correlated manner.

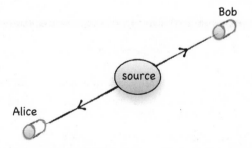

Figure 8.2 Experiment with a light source and two light detectors, which are observed to click apparently at random, but always simultaneously.

We are curious about the polarization properties of the light, so we each place a calcite crystal in our light beam to make polarization measurements. As described in Chapter 2, a calcite crystal splits a light beam into two polarized beams: one with polarization parallel to the crystal axis and the other with polarization perpendicular to the crystal's axis. FIGURE 8.3 shows the setup. Each of the crystals has an arrow marked on it indicating the crystal's axis. Behind each crystal is a pair of light detectors placed such that each polarized beam strikes one of them and possibly generates a click.

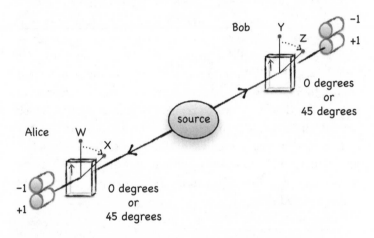

Figure 8.3 Experiment with a light source and two calcite crystals used to measure the polarization of light beams coming from a common source. On each side, the measurement scheme can be chosen as either zero degrees or forty-five degrees from vertical.

We design our experiment to mimic the correlation experiment with the two dancers described earlier, in which only one of two aspects was observed (arm or leg). You, Alice, agree to orient your crystal in one of two ways—either vertical (zero degrees from vertical) or at forty-five degrees from vertical. In the first case, you are measuring the polarization in the vertical/horizontal (V/H) scheme, and we agree to denote the measurement outcome by the symbol W. If the lower detector clicks, you will assign a value W = +1 to this outcome. If the upper detector clicks, you will assign a value W = −1 to this outcome. In the second case (forty-five degrees), you will be measuring polarization in the diagonal/antidiagonal (D/A) scheme, and we agree to denote the outcome by the symbol X. You will assign a value X = +1 if the lower detector clicks or a value X = −1 if the upper detector clicks. This naming of outcomes is analogous to AA, AL, and so on, as in the case of the dancers.

I, Bob, will do the same, using symbols Y and Z to denote the outcomes when measuring in two different schemes. In the case of a zero-degree crystal orientation for measuring in the H/V scheme, I will assign a value Y = +1 if the lower detector clicks or a value Y = −1 if the upper detector clicks. In the case of forty-five degrees for measuring in the D/A scheme, I will assign a value Z = +1 if the lower detector clicks or a value Z = −1 if the upper detector clicks.

As with arm and leg observations, there are four possible combinations of polarization measurements: WY, WZ, XY, and XZ. And for each of these combinations, there are four possible combined outcomes. For example, for WY, the outcomes could be (+1,+1), (+1,−1), (−1,+1), or (−1,−1). For the purposes of this example, I will choose a particular state of light to consider. Let me call it the *Bell State*, after John Bell. In Chapter 9, I explain more about this state and how it can be generated; for now, please accept that such a state can be generated when light is emitted by atoms, as I described earlier.

What kinds of correlations are observed in this case? When carrying out the experiment for light generated in the Bell State, we observe that if we both choose to measure H/V polarization, we always see opposite results: If you see +1, then I see −1. This means the H/V polarization measurements are perfectly negatively correlated. We also see perfect negative correlations if we both choose to measure D/A polarization. On the other hand, if we choose to measure using unlike schemes (for example, you use H/V and I use D/A), then we find the measurements are uncorrelated. These results are the same as discussed for observing two dancers whose arm (and leg) movements are negatively correlated, as summarized in TABLE 8.2. Apparently, there is "no mystery here."

But now, I, Bob, decide to change the two orientations at which I set my crystal axis for measuring the photon's polarization. In the next set of experiments, I choose randomly between 22.5 degrees and 67.5 degrees, as in FIGURE 8.4. In the case of 22.5 degrees, I will assign a value Y = +1 if the lower detector clicks or a value Y = −1 if the upper detector clicks. In the case of 67.5 degrees, I will assign a value Z = +1 if the lower detector clicks or a value Z = −1 if the upper detector clicks.

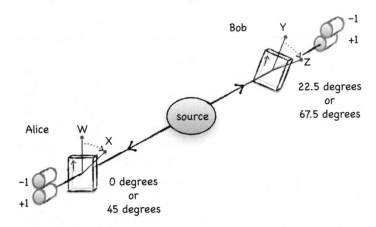

Figure 8.4 Similar to the previous figure, but here Bob switches between crystal angles of 22.5 degrees and 67.5 degrees.

You are probably wondering why we chose these particular crystal angles. In a real experiment, we would try out many different combinations of angles to get a more complete picture of the polarization properties. Now I am focusing on a particular combination that yields the most interesting results.

We, Alice and Bob, both make many observations in our separate laboratories while switching our crystal orientations randomly between our two crystal orientation possibilities. During these experiments, we have no communication, and to make sure there is no possibility of any mysterious, secret communication between the light beams or the detectors at our separate locations, we move our laboratories very far apart and make our measurements quickly. According to our understanding of Local Causality, there is no chance of anything in one laboratory affecting anything in the other.

Each of us sees only our own results for the time being. A portion of a long list of typical results is shown in TABLE 8.4.

Table 8.4 Lists of Outcomes of Photon Polarization Measurements W or X on Alice's Side, and Y or Z on Bob's Side

W	X	Y	Z	W × Y	W × Z	X × Y	X × Z
-1		1		-1			
-1			-1		1		
	1		-1				-1
	1	1				1	
1			1		1		
-1		-1		1			
1		1		1			
-1			-1		1		
	1		-1				-1
-1			-1		1		
	-1	1				-1	
1			-1		-1		
1		1		1			
1			-1		-1		
	-1	1				-1	
	-1		1				-1
-1			-1		1		
	-1	-1				1	
1		1				1	
1			-1		-1		
-1		1		-1			
-1			-1		1		
	1		-1				-1
	-1	-1				1	
Correlation =				0.7	0.7	0.7	-0.7

The four different correlations are given below each column of products of measured values.

Our own results look random to us before we compare with the other's results to look for correlations.

To spice things up, before we compare results, we decide to make a bet on the degree to which our observed results are correlated. That is, we bet on the value of the curious quantity Ave(Q) that will be obtained when we combine our results to calculate:

$$Ave\,(Q) = Ave(W \times Y) + Ave(W \times Z) + Ave(X \times Y) - Ave(X \times Z).$$

I bet that Ave(Q) can't possibly exceed 2, because I proved this mathematical fact earlier using only the completely reasonable hypotheses that define Local Realism.

Opposing my bet, you argue there is a weakness in my logic. You remind me that I am assuming it makes no difference that we measured only one outcome at a time (W or X for you, and Y or Z for me). That is, I am assuming the probabilities to observe +1 or −1 for each of these outcomes don't depend on whether we measure only one or measure both. But you point out that maybe it does matter. I counterargue that it doesn't matter, by analogy with the dancers experiment; their arm and leg positions *were* either up or down regardless of whether we observed them. Experiments with dancers would certainly bear this out, where Ave(Q) could never exceed 2. You agree that, for dancers, my argument is correct, but perhaps light is different. I counter, "No. Nothing can be so different that Local Realism is proved to be false. I bet my worldview on it!"

Now we both reveal our data lists and together calculate the correlations between our lists. We find the following results for the four correlations, which are the averages of the products: Ave(W × Y) = 0.7, Ave(W × Z) = 0.7, Ave(X × Y) = 0.7, and Ave(X × Z) = − 0.7. We notice the interesting minus sign on the correlation Ave(X × Z). Then, we combine these to compute the value of Ave(Q) using the usual formula just presented. We find Ave(Q) = 0.7 + 0.7 + 0.7 − (− 0.7) = 2.8, which certainly does exceed 2. I have lost my bet! My worldview of Local Realism is kaput. The experimentally proved failure of Local Realism is the main take-away message here.

Would all states of light emitted by the atoms produce this same result?

No. Only a specific type of quantum state of light, called a 'Bell State,' can produce such a result. For this reason, it is not likely that any scientists before 1900 would actually have stumbled on the precise experiment I just described. My point in telling the story this way is to emphasize that, if scientists had been very lucky, they *could* have stumbled on it without knowing any quantum physics, which wasn't discovered until after 1900.

In modern times, we think of light as comprised of photons. The experiment I described involves pairs of photons described by the Bell State, which is an example of an *entangled quantum state*. Alain Aspect wrote in 1999, "We must conclude that an entangled photon pair is a non-separable object; that is, it is impossible to assign individual local properties (local physical reality) to each photon. In some sense, both photons keep in contact through space and time."[3]

This view seems to be a rather drastic reinterpretation of the concept of physical reality, if we grant that we believe Einstein's Principle of Local Causality, which prevents the photons from communicating information between them during the experiment. Yet the correlations leading to Ave(Q) = 2.8 do occur in the real world despite there being no communication between the two distant laboratories. And we know that in the absence of such communication, any model of Nature that assumes each photon carries preassigned properties or instructions prohibits Ave(Q) from being greater than 2.

Are there possible flaws or loopholes in our arguments?

"Wait!," you might say. "Before we recklessly throw out our beloved concept of local reality, we should examine critically how the experiments were done. Maybe we can save Local Realism by pointing out a flaw in the assumptions behind the analysis of the experiments." The two potential flaws that had not been overcome simultaneously until the landmark experiments in 2015 were (1) efficiency of detection and (2) quickness of setting the measurement schemes.

In the early experiments, the problem of efficiency of detection was that not all photons created in the source and sent to the separated measurement stations were captured and detected. Some were lost or escaped detection. Skeptics could then argue that a grand conspiracy of Nature could, in principle, be at work, which secretly allowed only certain correlated photon pairs to be detected. This would skew the measured correlations, obscuring the true correlation values that would be measured if all photons were captured and measured. Developing the technology that allowed experimenters to capture and measure nearly all photons was a formidable task and took several decades to accomplish.

The second flaw that needed overcoming was being able to ensure the measurement schemes being used for each measurement (W or X for Alice and Y or Z for Bob) were chosen independently of each other, and that they were set after the photons had departed from the source and at a time just before each measurement. The quick setting of measurement schemes ensures no communication between laboratories is possible during the time each measurement is carried out. If the settings are not chosen independently and well after the photons had departed from the source, then it is impossible to rule out the possibility of a conspiracy of Nature in which the photons 'knew' in advance what the measurement settings would be. The photons could then, in principle, have had advance instructions 'programmed' into them to yield measurement correlations that would exceed that allowed by the

Bell Relation. That is, Nature could 'fool' us into thinking Local Realism is not valid, when actually it is.

What experiments have overcome the potential flaws?

The 1982 experiment of Aspect and coworkers, mentioned earlier, was the first to attempt setting the photon measurement schemes after the photons had been created and were traveling through the laboratory on their way to the detectors. The calcite crystal orientations that set the measurement schemes were varied periodically and so rapidly that the scheme had changed from its original setting by the time a photon arrived at the detector. The Bell relation, $\text{Ave}(Q) \leq 2$, was violated by the observed data, giving strong evidence against the worldview of Local Realism.

Nevertheless, some scientists argued that the measurement schemes shouldn't be varied periodically or regularly, as in the Aspect experiment, but randomly. A random variation would rule out any unknown quirk of Nature that could allow the measurement schemes to be 'known' in advance by the photons as they were being created, allowing them to coordinate in advance their behavior when they reached the detectors. Subsequent experiments by other groups redid the experiments using rapid and random setting of the measurement schemes, and again the Bell Relation was violated. But still, the detectors they used were not efficient enough to rule out the other potential flaw or loophole mentioned earlier: that the detectors somehow were able to select only certain pairs of photons in a way that would lead to an apparent violation of the Bell relation, and 'trick' the scientists into thinking they had ruled out Local Realism by their experiment.

Finally, in 2015 three independent laboratories carried out experiments in which both of the previously described technicalities or loopholes were eliminated.[4] One group at the National Institute of Standards and Technology in Boulder, Colorado, and another group at the Institute for Quantum

Optics and Quantum Information at the University of Vienna, measured photon polarizations in experiments much like I just described. A third group at Delft University of Technology carried out a different experiment; they created and detected an entangled state of two *electrons* that were located in two different diamond crystals separated by 1.3 kilometers. In all three, new, highly efficient detectors were used, and machines were programmed to generate random measurement settings very rapidly, to rule out the loophole explanations described earlier. Again, the Bell Relation was violated.

How can we be sure the measurement settings are independent?

A third loophole, which may seem to weaken the conclusions of the just-described experiments, is the possibility that some unknown aspect of Nature 'conspires' *before* each photon pair is created to 'force' the computers that set the measurement schemes in the distant laboratories to do so in particular ways. Given enough time in advance of the photons' creation, such a phenomenon would not be in conflict with relativity's upper speed limit for communicating information. This kind of unknown 'conspiracy of Nature' could then cause the settings to be such that the measurement correlations violate the Bell Relation even though Nature does actually behave according to Local Realism.

To eliminate this as a hypothetical possibility that would 'trick' the experimenters into misinterpreting their results, an experiment was carried out in 2017 by Anton Zeilinger and collaborators at the University of Vienna. They used light received from two distant stars to set the measurement schemes randomly and quickly at each detector. Because the light received had been emitted by the stars more than five hundred years ago, the experiment ruled out the possibility that an unknown aspect of Nature 'conspires' *just before* each photon pair is created to 'force' the measurement schemes to be set in particular ways. This experiment gave the most stringent test possible of Local Realism, which again failed to explain the observed results. With the technicalities, or loopholes, finally overcome, physicists could say with near certainty that Local Realism is untenable as a view of the physical world.[5]

What did John Bell make of the results of such experiments?

Albert Einstein, one of the founders of quantum theory, died before the Bell-test experiments were carried out. He was probably the greatest thinker on this matter who went to his grave believing in Local Realism. Recall that John Bell was responsible for the line of thinking that led to the end of Local Realism as a valid worldview. He lived to see the results of Aspect's experiments, in which photons emitted in pairs by an atom were observed to violate the Bell Relation. In his own words:

> For me, it is so reasonable to assume that the photons in those experiments carry with them programs, which have been correlated in advance, telling them how to behave. This is so rational that I think that when Einstein saw that, and the others refused to see it, he was the rational man. The other people, although history has justified them, were burying their heads in the sand. I feel that Einstein's intellectual superiority over Bohr, in this instance, was enormous; a vast gulf between the man who saw clearly what was needed, and the obscurantist. So for me, it is a pity that Einstein's idea doesn't work. The reasonable thing just doesn't work.[6]

Or, as I like to say, Einstein was wrong, but for all the 'right' reasons. The notion of realism is strongly embedded in the human psyche. The idea conforms to common sense, as the examples of socks and dancers show. Furthermore, no successful theory before quantum mechanics had the ability to predict the existence of correlations in cases in which preexisting properties or instruction sets were absent. Finally, if the classical physics ideas about properties and instruction sets are false, then it is very hard to think of any intuitive model that makes understandable the correlations observed in the Bell-test experiments.

Does the breakdown of Local Realism mean we must abandon classical intuition and classical physics altogether?

No. We know that, in a strict sense, classical physics is not the most correct description of Nature, but still, in the vast majority of everyday experiences, it is the best description and very useful. The cases in which the ideas of classical realism fail are few and far between. As Michael Nielsen, a pioneer in quantum information theory wrote:

> For most practical everyday purposes, we can treat a coin as knowing whether it is heads or tails, and a cat as knowing whether it is alive or dead. Although these beliefs are not correct at some fundamental level, in most practical situations they work extremely well. It's only in extraordinary circumstances quite outside everyday life that this way of thinking could ever lead you astray.[7]

That's a relief, at least.

Should we abandon Local Causality or Local Realism, or both?

There is not a wide consensus among physicists on the answer to this question. Many physicists (including me) take the view that because relativity is so well established, and it prohibits any causal influence from traveling faster than light, a sensible physics theory should respect the Principle of Local Causality. Quantum theory itself does not predict any violations of Local

Causality. This seems to imply that we should retain Local Causality in our worldview, and this would mean we have to give up realism. As John Bell pointed out earlier, he and Einstein were quite unhappy about this prospect.

In favor of abandoning realism, note that quantum theory does not rest on the assumption that entities have definite, if unknown, intrinsic properties before being measured, independent of the way in which they are measured. In fact, a famous mathematical proof called the Kochen-Specker Theorem shows that, under the assumption that quantum theory is the true theory of Nature, physical entities cannot and do not have definite, preexisting properties before their being measured.

Therefore, it seems consistent with all experiments and with quantum theory to say that Local Causality should be kept and Local Realism abandoned. Nevertheless, other scientists might prefer to give up Local Causality, and many are 'agnostic' on these questions. What we *can* say confidently is that the correlations seem to have a nonlocal nature to them. This fact is still puzzling to most physicists, who hope that future studies will help clarify it.

Figure Notes

Figure 8.1 is from John Bell, "Bertlmann's Socks and the Nature of Reality," *Journal de Physique Colloques* 42 (1981): C2-41–C2-62; figure, C2-42.

Notes

1 The quote is from John S. Bell, "Bertlmann's Socks and the Nature of Reality," in J. S. Bell, *Speakable and Unspeakable in Quantum Mechanics* (Cambridge: Cambridge University Press, 1987), 139.

2 Einstein predicted, in 1916, that violent interactions involving massive objects could 'shake off' gravitational waves that would travel across the universe at the speed of light—and no faster. In

2015, such gravitational waves were detected for the first time, by the LIGO Collaboration.

3 The quote is from Alain Aspect, "Bell's Inequality Test: More Ideal Than Ever," *Nature* 398 (1999): 189–190; quote, 190.

4 For a readable description of the three 2015 Bell-test experiments, see Alain Aspect, "Viewpoint: Closing the Door on Einstein and Bohr's Quantum Debate," *Physics*, American Physical Society, http://physics.aps.org/articles/v8/123.

5 Some scientists point out that experiments such as the one using starlight to set the measurement schemes in the Bell-test merely push the possibility of unknown controlling factors in Nature to earlier epochs, but don't rule them out entirely. Taken to the limit, the epoch could be pushed all the way back to the Big Bang, implying that everything that happens in the Universe, down to the smallest event, is predetermined. Most scientists view this idea, called superdeterminism, as an unsatisfactory view of Nature.

6 The quote is from Jeremy Bernstein, *Quantum Profiles* (Princeton, NJ: Princeton University Press, 1991), 84.

7 The quote is from the blog entry of Michael Nielsen titled "Why the World Needs Quantum Mechanics," August 4, 2008, http://michaelnielsen.org/blog/why-the-world-needs-quantum-mechanics/.

9

QUANTUM ENTANGLEMENT AND TELEPORTATION

What is quantum entanglement?

Quantum entanglement is a characteristic of special quantum states describing two or more quantum entities, such as photons or electrons. It represents a form of measurement correlation that occurs only in quantum physics. Recall from Chapter 8 that the Bell Relation, Ave(Q) ≤ 2, characterizes limits on all possible classically correlated measurements. The Bell Relation can be violated by some quantum entities, but only if the entities are prepared in an entangled quantum state. In this sense, entanglement refers to a type of measurement correlation that goes beyond what is possible in classical physics. The experiments carried out during the late 1970s and 1980s were the first to reveal violations of the Bell Relation, and by this observation were the first to detect entanglement in the state of separated quantum objects, such as pairs of photons. Bell-test experiments are, recall, a test of the worldview called Local Realism. Thus, quantum entanglement is involved in the breakdown of Local Realism.

The take-away message of the Bell Relation-violating experiments is that individual quantum entities do not carry preexisting properties or predetermined instructions for their behavior on measurement. This statement followed plausibly from analyzing the measurement correlations observed in experiments, while making the well-founded assumption that two separated objects cannot communicate or influence each other faster than the speed of light. But, at the same time, for certain choices of measurement schemes on Alice's and Bob's sides, perfect correlations were observed. For the Bell State, if both choose to measure using horizontal/vertical polarization schemes, they always see opposite results; if Alice sees H, then Bob always sees V, and vice versa. That is, the outcomes of H/V polarization measurements are perfectly negatively correlated. In fact, as long as Alice and Bob choose the same measurement scheme, no matter at what orientation, they will observe perfectly negatively correlated outcomes.

It probably seems strange—perhaps astounding—that although individual quantum entities do not carry preexisting properties or predetermined instructions, and they do not communicate, they can still display perfect correlations upon measurement. The good news is that quantum theory does represent such a result perfectly.

An application of entanglement is quantum teleportation— a technique that enables transmitting the state of one quantum object to another at a distant location. I explain how this works at the end of this chapter. The most important application of quantum entanglement is probably quantum computing, which is aimed at building computers that process information in a fundamentally quantum way, instead of in a classical way, as is the case for all computers today. Chapter 10 explores these ideas.

How do we represent the state of a composite entity?

A 'composite entity' is a physical entity composed of two or more parts or objects, although these parts may be well separated. A good example is a pair of photons. Before we discuss entangled states, let's consider how to represent the quantum state of a composite entity.

Recall that a quantum state is as complete a description as can be given about a quantum entity or object. If we know the quantum state fully, there really is no more information to be had. Yet, even knowing the state perfectly, we cannot predict with certainty the outcome of most measurements. Nature has random behavior at its deepest level.

We discussed in Chapter 4 how to represent the state of a single quantum entity by a state arrow. For a photon we can represent a state of vertical polarization (V) as a vertical arrow, which I put into parentheses as (↑). Likewise, a state of horizontal polarization (H) is represented as (→). Diagonal (D) and antidiagonal (A) polarization states are represented pictorially as (↗) and (↖), respectively.

We also stressed that a quantum object can be prepared in a state that is a superposition of two distinct states. For example, the diagonal polarization state is a superposition of vertical and horizontal, which I illustrate in FIGURE 9.1(i) and write symbolically as a sum of vertical and horizontal state arrows: (↗) = (↑) + (→). In quantum theory the plus sign (+) stands for 'in superposition with.' (To jog your memory about how arrows sum together, recall that the northeast direction is composed of equal parts of north and east directions.) Likewise, antidiagonal polarization is a sum of these two states in which the horizontal component is flipped to the opposite direction—that is, (↖) = (↑) + (←). In FIGURE 9.1(ii), the minus sign on –H indicates this flipped horizontal component (←).

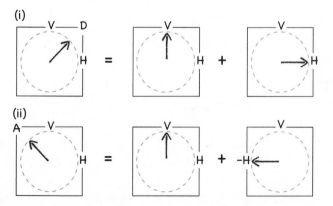

Figure 9.1 Illustrating state arrow superposition for the polarization state of a single photon. (i) Diagonal polarization (↗) = (↑) + (→). (ii): Antidiagonal polarization (↖) =(↑) + (←).

How do we represent an entangled state of a pair of photons?

An *entangled state* is a superposition of two or more quantum possibilities for a composite entity. Such states arise in quantum theory because the concept of state superposition applies not only to individual objects, but also to composite entities treated as a whole. An entangled state is said to represent or contain *entanglement*.

Consider a pair of photons. If I want to specify the states of Alice's photon A and Bob's photon B, I need to write a state arrow for each. For example, if photon A has vertical polarization and photon B has horizontal polarization, I can represent this composite state as {(↑) *and* (→)} or, for short, (↑)&(→). Here, the arrow on the left refers to A and the arrow on the right refers to B. This state is not an entangled state, because it specifies that each photon has a specific, known polarization state that is specified independently of the other.

An example of an entangled state of two photons is the Bell State, which we need to create for performing the experiments in which the Bell Relation is violated. It is represented pictorially in FIGURE 9.2 and symbolically by

$$(↑)\&(→) + (←)\&(↑).$$

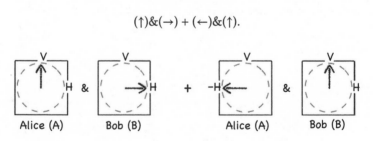

Alice (A)   Bob (B)   Alice (A)   Bob (B)

Figure 9.2 The Bell State for a composite entity consisting of a pair of photons.

This state is a superposition of two quantum possibilities for the two photons considered as a composite entity. The state can be said in words as {photon A is polarized vertically and photon B is polarized horizontally} in superposition with {photon A is polarized vertically and photon B is polarized horizontally}.

This means that if we were to measure the two photons using calcite crystals to distinguish between H and V polarizations, we would observe as outcomes one of the two combined possibilities H&V or V&H. That is, the results of measuring the polarization of the photons will be perfectly negatively correlated, just as observed in the Bell-test experiments.

If the A photon were sent to Alice's laboratory and the B photon to Bob's laboratory, then no matter how large the distance between the laboratories, the same correlation would appear upon the measurements being carried out. You can't predict which possibility you will observe, but in each possibility there will be a definite relationship between the measurements of the two photons.

You might think, "What's the big deal? I could observe the same behavior if I have two balls, one red and one blue, and send one to Alice and one to Bob. If Bob looks at his received ball and sees red, then it's obvious to him that Alice has received the blue one. That is simple common sense." No, it's not that simple. The results of the Bell Relation experiments discussed in Chapter 8 show that the correlations observed in the measured polarizations of photon pairs in the entangled Bell State are not simply analogous to classical correlations such as those of colored balls. Specifically, Alice and Bob can measure their photons using various complementary measurement schemes: H/V or D/A, and so on, and they can observe correlations that violate the Bell Relation. There is something much deeper and more surprising going on, which entangled states do describe fully.

How can we make the Bell State for a photon pair?

A method for generating the Bell State is illustrated in FIGURE 9.3. Although this method is not the one used in the first experiments carried out during the 1970s with the intention of violating the Bell Relation, it illustrates the key points about entanglement.

Consider separate photon sources, labeled A and B, that can each emit a single photon. Photon A is always V-polarized. Photon B may be either H-polarized or V-polarized, and we control which. The photons pass through a device, labeled U, that is built to have the following behaviors: The device passes photon B unchanged regardless of whether it is H-pol or V-pol, but the device may change the polarization state of photon A, depending on the polarization of photon B. As shown in FIGURE 9.3(i), if photon B is H-pol, then photon A passes through unchanged, so the state emerging at the right is the same as before—that is, $(\uparrow)\&(\rightarrow)$. On the other hand, if photon B is V-pol, as in FIGURE 9.3(ii), then photon A is changed to negative H-pol, so the state that emerges at the right is $(\leftarrow)\&(\uparrow)$.

Now consider what state will be generated if the B photon is diagonal-polarized—that is, $(\nearrow)$—as in FIGURE 9.3(iii). The combined state of photons A and B is then $(\uparrow)\&(\nearrow)$. We know

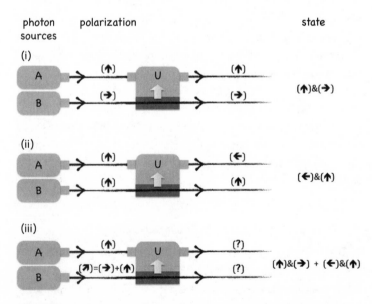

Figure 9.3 A Bell State generator. The state of the B photon controls the final state of the A photon, as indicated by the light-gray arrow.

that the diagonal state of photon B is an equal superposition of the two possibilities H-pol and V-pol—that is, $(\nearrow) = (\rightarrow) + (\uparrow)$. Furthermore, the experimental setup is built so the entire process is a unitary one (thus the label U). That is, no intermediate measurement is performed on either photon that would yield a measurement outcome H-pol or V-pol. Therefore, the two possibilities H and V for the B photon equally influence the possible changing of the A photon. The resulting state describing the two photons together as a whole is expressed as

$$(\uparrow)\&(\rightarrow) + (\leftarrow)\&(\uparrow).$$

Again, the + symbol means 'in quantum superposition with.' This is the desired Bell State.

How does the entangled Bell State violate Local Realism?

Say two photons are prepared in the Bell State. Then, if Bob and Alice both measure their photons using the H/V scheme, using calcite crystals and detectors as described in Chapter 2, they will observe opposite results. If Bob observes H-pol, corresponding to the $(\uparrow)\&(\rightarrow)$ part of the entangled state, Alice will observe V-pol. But, if Bob observes V-pol, corresponding to the $(\leftarrow)\&(\uparrow)$ part of the entangled state, Alice will observe H-pol.

On the other hand, if Bob uses the H/V scheme for measurement and Alice uses the D/A measurement scheme, so she is analyzing for diagonal or antidiagonal polarization, then there will be no correlation between their measurement outcomes. Regardless of whether Bob observes H or V for his photon, Alice will have a fifty-percent probability of observing either D or A. This can be understood by recalling that, for example, the vertical polarization state is an equal superposition of the two possibilities D-pol and A-pol—that is, $(\uparrow) = (\nearrow) + (\nwarrow)$.

The correlations discussed so far, using only the H/V scheme or D/A scheme, do not violate the Bell Relation. In such cases, the correlations behave like classical correlations. In contrast, as described in Chapter 8, if Alice continues using the H/V (or D/A) measurement scheme, and Bob uses a 22.5-degree (or 67.5-degree) measurement scheme, then the measured correlations will yield a value of 2.8 for the average of the 'curious quantity' Q. This value exceeds 2, thus violating the Bell Relation. This result cannot be explained as resulting from a classical correlation; that is the whole point of the Bell Relation. It can be shown that quantum theory correctly predicts the value for the average of Q—namely, 2.8—precisely in agreement with the experiment. The proof is rather involved, so I won't show it. In contrast, any theory based on a worldview of Local Realism cannot predict the results correctly. Clearly, quantum theory correctly incorporates drastically different physics than any theory based on Local Realism.

What can you know about the constituents of a quantum composite object?

Erwin Schrödinger, who introduced entanglement into quantum theory in 1935, wrote, "If two separated bodies, each by itself known maximally, enter a situation in which they influence each other, and separate again, then there occurs ... *entanglement* of our knowledge of the two bodies."[1]

Recall the essence of 'state' in quantum physics: If you know the quantum state of a composite entity as a whole, that is, the complete description of it, you know everything there is to know about the entity. But what do we know about the constituents of the composite entity? To quote physicist Leonard Susskind,

> "You can know everything there is to know about a composite system [or entity], and yet *not* know everything about the individual constituents."[2]

This is a remarkable statement. There is nothing like this situation in a classical worldview. When thinking in a classical physics way, we naturally think that to know everything about a composite entity, such as a pair of objects, we would need to know everything about each of its constituents. For quantum composite objects or entities, this is not the case, nor is it even possible in general. You can never know everything about a quantum object, in the sense that you cannot predict with certainty the outcomes of every possible measurement on it. The impossibility of knowing everything applies also to composite entities; the most that can be known is contained in the quantum state, which may be entangled, as in the previous example.

This fact of Nature perhaps expresses the deepest meaning of entangled states. And it is what enables the experiments described earlier to violate the Bell Relation. To summarize:

An *entangled state* of a composite entity is a state that provides a maximally complete quantum description of the whole entity, but nevertheless cannot be divided into maximally complete quantum descriptions of its constituent parts.

What does it mean in practice to know everything there is to know about a composite quantum entity?

For quantum entities we cannot know with certainty what will be the measurement outcomes for all measurable quantities, such as polarization, position, path, and so on. So we cannot know everything in that sense. But if we do know the quantum state that describes the entity, there is always at least one quantity whose measurement outcome we can predict with one-hundred-percent confidence.

For example, consider a single photon described by a horizontal (H) polarization state. If we measure this photon using a calcite crystal oriented so it splits the incoming light beam into horizontal and vertical components, we can predict with full confidence that it will go into the H-pol beam and be detected. That is, we have predicted its measurement outcome with certainty.

In quantum physics, being able to predict a single particular measurement outcome with certainty defines what we mean by knowing everything there is to know. Knowing in advance the outcome or answer to this single question provides more information than might be obvious at first glance. This knowledge confirms for us what the state of the quantum entity is. Knowing the state, in turn, allows us to calculate *probabilities* for any other polarization measurement we might wish to perform on any photon prepared in this same state. For example, if we know a photon has H polarization as its quantum state (which we just verified by measurement), we can predict that a measurement in the diagonal/antidiagonal measurement scheme will yield the two possible outcomes, D or A, with a fifty-percent probability for each.

For a composite entity, such as a photon pair, the same statement holds. If we know the quantum state of the composite entity, there is always one measurement we can perform with an outcome we can predict with one-hundred-percent confidence. For example, let's say we know the pair has been prepared so its state is the Bell State, as described earlier. Then, we can devise a measurement scheme with an outcome we can predict with certainty.

Figure 9.4 A Bell State verifier. To perform a measurement for which we can predict the outcome with certainty, send the photon pair from right to left to undo the entangling operation, then measure the polarizations of each photon.

The experimental scheme for performing this measurement is shown in FIGURE 9.4. It uses the reverse of the process we used to create the Bell State. By running the photon pair back through the same device, we can undo the entangling process and recreate the original photons' combined state (↑)&(↗). Then, it is easy to perform measurements on each photon that will yield perfectly predictable outcomes. For the photon in the upper beam, we use a calcite crystal to measure its polarization in the H/V scheme, for which we can be certain the result will be V. For the photon in the lower beam, we use a rotated calcite crystal to measure its polarization in the D/A scheme, for which we can be certain the result will be D. The fact that we can predict these two outcomes with certainty is the sense in which we know everything there is to know about the composite photon pair.

For this measurement scheme to work, both photons must enter the same measurement device, which means they must interact locally with the device and with each other. This fact leads to an important realization: If Bob, for example, takes one of the photons and refuses to share it with Alice, then Alice is unable to predict what polarization her photon will display if measured. In fact, for the case of the Bell State, Alice's photon, when measured on its own, can be observed to have any polarization with equal probability. She has no information at all about the polarization state of her photon by itself. If Bob informs Alice, say by telephone, "I'm telling you that the photon pair is in the Bell State, so we know everything there is to know," Alice will retort, "That information is of no help to me. As long as you continue to hoard your photon, I can make no reliable, certain prediction about my photon by itself."

What can we accomplish using entanglement that we couldn't without it?

Besides underpinning the counterintuitive correlations observed in experiments that violate the Bell Relation, entanglement also enables interesting operations or processes, some of which have useful applications in communications and computing technologies. In this sense entanglement can be thought of as a useful resource in a similar way that energy is a resource.

One potential use of entanglement is for increasing the ability of computers to solve certain computational problems in a reasonable amount of time. Quantum computers are discussed in Chapter 10.

Another unexpected use of entanglement is for 'teleporting' a quantum state from one location to another. This does not mean teleporting a physical object from place to place; that is impossible, as far as we know. And we are not simply transferring knowledge of the state from one place to another; that could be done using a telephone. The goal of quantum teleportation is to move or transmit a quantum state of an object at one location to an object at another location without actually knowing the state or sending the object itself.

If you were to know perfectly the quantum state of a photon, you could transmit its state simply by placing a phone call or sending an e-mail to a distant friend describing in detail this state. Then the friend, named Bob, could prepare a new photon in that state. No teleportation needed. The more interesting case is if someone hands you a quantum object and doesn't tell you the state but asks you to teleport it to Bob anyway. This might seem impossible; as I pointed out in Chapter 2, you can't make a copy of the state—a principle of quantum physics called the No Cloning Principle—and you can't determine the state by making measurements on that single object. We discussed both of these points in Chapter 2. So you don't possess the information needed to instruct Bob how to recreate or mimic the state in his laboratory.

How does entanglement enable quantum state teleportation?

Entanglement offers a way to accomplish teleportation of a quantum state without knowing that state. The method is illustrated in FIGURE 9.5. Say Professor Xavier has sent Alice a photon for which he knows perfectly the polarization state but about which she knows nothing. Call this photon X and the polarization state of this photon (Ψ). An example of this state is illustrated by the state arrow in the inset in FIGURE 9.5. This angle could point anywhere in the H/V state diagram. Only Professor Xavier knows this state.

Alice wants to teleport the photon's state to Bob without ever knowing what the state is herself. Before Alice can begin the teleportation procedure, she and Bob need to arrange that they each have one of two photons, which I will naturally call A and B, and which are prepared in an entangled Bell State. This pair of photons is prepared in the Source (see FIGURE 9.5), and one-half of the entangled pair is transmitted to Alice and to Bob (which may take some time, so advance ordering is recommended!). After Alice receives photon A and Bob receives photon B, they are ready to teleport the state of photon X.

First, Alice passes photons A and X together through the Bell State verifier shown earlier in FIGURE 9.4. As before, this device flips (H⟷V) the polarization of the X photon if the A photon has V-pol, but leaves the X photon unchanged if the A photon has H-pol. Because the A photon is described by a

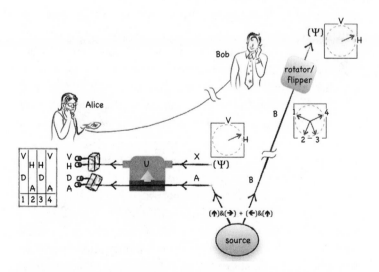

Figure 9.5 Quantum state teleportation. The state (Ψ) of photon X is teleported from Alice to Bob without actually sending photon X to Bob. (The inset artwork of Alice and Bob was created by Dawn Hudson and is used with permission.)

superposition state containing both V and H, the device has the effect of creating an entangled state of photons A and X. Because photon A was already entangled with photon B, now all three photons are entangled. Bob knows nothing at this point except that photons A and B were entangled initially.

Second, Alice measures the X photon in the H/V measurement scheme and the A photon in the D/A scheme. As shown in the table in FIGURE 9.5, there are four possible combinations of outcomes for these measurements: VD, HA, HD, and VA. These outcome combinations are labeled 1, 2, 3, and 4 in FIGURE 9.5. Quantum theory shows that, for each of these outcomes, the B photon on Bob's side will be left in a particular state. This state would be perfectly known to Professor Xavier if he were monitoring the proceedings. (The details of the proof are a little complicated, so they aren't given.) The state arrows corresponding to these four states are shown on Bob's side, although he is still ignorant of what is going on. Alice now knows how the state of the B photon is related to the original state given to her by the professor, although she does not know its actual state.

Third, Alice calls Bob on the telephone and informs him which of the four outcomes she observed. Now, because Bob knows the theory behind teleportation, he knows how his photon's state is related to the original state given to Alice. If Alice reports outcome 4, Bob is assured that photon B is described by the same state as the original photon X, although he doesn't know the actual state. If Alice reports outcome 1, the state of the B photon is not the same as the original X photon's state. Nevertheless, Bob is assured that photon B will be correctly described by the X photon's state after he simply flips the polarization arrow from left to right. For the other two cases, outcomes 2 and 3, there are also simple alterations of the photon's polarization state that will cause the B photon to be identical to the X photon's state. He implements the required operation using the device labeled the rotator/flipper in FIGURE 9.5. It consists of optical materials that alter light polarization in a known way. Bob has to set the action of the device according to the information he receives from Alice via the phone call. After such operations are carried out, quantum theory guarantees that the original X photon's state now describes perfectly the B photon.

The X photon is no longer described by its original state. Thus, the state has been transferred from one photon to another. Teleportation of the quantum state has been accomplished, without Alice or Bob ever knowing the state itself and without leaving a copy of the state in Alice's laboratory. Experiments of this type have been carried out successfully since the 1990s.

Does what happens on Alice's side affect what happens on Bob's side?

No. In a situation where Alice and Bob each have one-half of an entangled photon pair, no matter what Alice does with her detection setup, she cannot affect or change the state of the photon on Bob's side. In a sense, this situation can be thought of as analogous to the example with the red and blue balls mentioned earlier. If Alice and Bob each had a ball but were ignorant of which they had, then by Alice merely determining the color, or state, of her ball, she would know the color or state of Bob's ball. There is no physical mechanism at work, only information becoming known to Alice.

Admittedly, in quantum physics the explanation is more subtle: a quantum state is not in one-to-one correspondence with predetermined properties. Nevertheless, in quantum theory, there are definite relationships between different possibilities contained in a superposition state. These relationships replace the simple logic that we used to understand the colored balls situation. For example, if Alice observed outcome number 3, then she knows from quantum theory that the state arrow labeled 3 describes Bob's photon perfectly.

Note that if, before using the rotator/flipper, Bob were to measure his photon using an H/V calcite crystal, he would observe equal numbers of horizontal and vertical outcomes. Information acquired by Alice would in no way change Bob's observations. Quantum theory merely predicts the correlations that will exist between measurements made by Bob and by Alice.

The fact that any local, realistic theory fails to predict the correlations observed in Bell Relation-violating experiments has led some people to think, wrongly, that there must be a cause-and-effect between Alice's and Bob's sides. Quantum theory—the only theory we know that predicts the Bell-type correlations correctly—does not contain any such cause and effect between distant events. It has no need for such a mechanism.

Is quantum teleportation instantaneous?

No. Quantum teleportation is not instantaneous. The needed entangled photon pair A, B can be provided long before the teleportation procedure is carried out. It doesn't matter how long it takes to provide the pair, as long as it is done well in advance. Then, the teleportation can take place in as short a time it takes for Alice to make her measurement and inform Bob of the result, via phone or, even faster, by direct radio link. Although Alice knows the result of her measurements instantly when she views them, and she knows them before Bob does, this information is of no use to Bob until he receives the information, which cannot travel to him any faster than the speed of light.

Can a human be teleported?

Although most physicists believe that all material objects are subject to the laws of quantum physics, and thus could, in principle, be teleported, it is highly unlikely it will ever be possible to teleport the quantum state of a human from one place to another. Say Alice and Bob wanted to teleport a person named Sulu. Imagine what would be required to do so. Recall that matter is not teleported; only a quantum state is teleported. Therefore, at the destination end, Bob would have to supply a pile of raw materials, atoms of carbon, oxygen, nitrogen, and so on, arranged in a manner ready to receive the incoming state. Alice and Bob would need to have at their disposal around 10^{26} entangled pairs of atoms to use as the needed resource, one for each atom in Sulu's body. Each of those entangled atoms would have to be put through a Bell State verifier, along with each corresponding atom taken from Sulu's deconstructed body. Then, Alice would need to place a phone call to tell Bob the results of each of the 10^{26} measurements. And Bob would need to receive that information and use a state rotator/flipper on each of the atoms in his pile of raw materials, while at the same time doing the unimaginable task of assembling Sulu's new atoms in the correct arrangement.

If the information could be transmitted at one terabyte per second, which is beyond any current capability, it would still take more than a million years for all of it to be sent to Bob. Beyond the need to wait so long, the ability to deal with and control these huge numbers of atoms seems beyond belief, even in principle.

What is quantum state teleportation good for?

If there is a need to transfer a quantum state from one location to another as quickly as possible (although not instantaneously!), teleportation is the way to go. Currently, there are limited situations in which this has a practical use. Within the next ten years or so, as quantum technologies become better developed and prevalent, the need for practical uses will increase. One future use will likely be for passing quantum information, or data, between quantum computers. These are machines that scientists are now attempting to construct, which would process so-called quantum information. Such machines are explored further in the next chapter.

Notes

1 The quote is from Erwin Schrödinger and it appears in John D. Trimmer, "The Present Situation in Quantum Mechanics: A Translation of Schrödinger's Cat Paradox Paper," *Proceedings of the American Philosophical Society* 124 (1980), 323–338; quote, 332.

2 The quote is from Leonard Susskind and it appears in Peter Byrne, "Bad Boy of Physics," *Scientific American* 305 (2011), 80–83; quote, 82.

10

APPLICATION: QUANTUM COMPUTING

Is information physical?

Computers process information. Computer scientist and physicist Rolf Landauer was a proponent of the idea that information is an aspect of the physical world. He elaborated this as follows:

> Information is not a disembodied abstract entity; it is always tied to a physical representation. It is represented by engraving on a stone tablet, a [magnetic] spin, [an electric] charge, a hole in a punched card, a mark on paper, or some other equivalent. This ties the handling of information to all the possibilities and restrictions of our real physical word, its laws of physics and its storehouse of available parts.[1]

If "information is physical," as Landauer has said, then it would seem necessary to treat it quantum mechanically. That is, the physical means by which information is stored and processed by computers should be considered using quantum theory. Before discussing quantum computers, it helps to understand computing in general.

What is a computer?

A computer is a machine that receives and stores information input, processes that information according to a programmable sequence of steps, and creates a resulting information output. The word 'computer' was first used during the 1600s to refer to persons who perform calculations or computations, and now refers to machines that compute. Machines that perform computations can be classified roughly into four kinds:

1. **Mechanical classical physics calculating machines.** These machines work using moving parts, including levers and gears, to perform computations. Typically they are not programmable, but always perform the same task, such as addition of numbers. An example is the Burroughs adding machine of 1905.

2. **Electromechanical classical physics fully programmable calculating machines.** These machines work using moving parts controlled by electronics. They process information stored as digital bits represented by the positions of large numbers of electromechanical switches. The first such machine was built by Konrad Zuse in 1941 in wartime Germany. Their programmability makes them able, in principle, to solve any problem that can be stated and solved using algebra. In this sense, they were the first 'universal' computers.

3. **All-electronic, hybrid quantum–classical–physics computers.** These fully programmable, universal computing machines have no moving mechanical parts and work using electronic circuits. The first to be built was the ENIAC, designed by John Mauchly and J. Presper Eckert of the University of Pennsylvania, in 1946. The physical principles describing the motion of electrons in these circuits are rooted in quantum physics. But, because there are no superposition states or entangled states involving electrons in different circuit components (capacitors, transistors, and so on), classical physics describes adequately

the manner in which electrons represent information. Therefore, we call these machines—essentially every computer in operation today—'classical computers.'

4. **Quantum computers.** If ever constructed successfully, these machines will work using intrinsically quantum physics principles. Information will be represented by quantum states of individual electrons or other elementary quantum objects, and entangled states will exist involving electrons in different circuit components. Such computers are predicted to be capable of solving certain kinds of problems far more quickly than any present-day classical computer can.

How do computers work?

As we discussed briefly in Chapter 3, computers store and manipulate information using a *binary* language with an alphabet that consists of only two symbols: 0 and 1. Each 1 or 0 symbol is called a *bit*, short for binary digit, because it can take on one of two possible values. A page of text, such as the one you are reading, is represented in a computer file as a long string of numbers. Every letter is represented by a binary code. For example, 'A' becomes 01000001, 'B' becomes 01000010, and so on.

In a typical computer, each bit is represented by the number of electrons stored in a tiny device called a capacitor. We can think of a capacitor as a box that holds a certain number of electrons, kind of like a bulk-grain bin at the food store holding a certain amount of rice. Each capacitor is called a memory cell. For example, such a capacitor might have a maximum capacity of one thousand electrons. If a capacitor is full or almost full of electrons, we say it represents a bit value of 1. If the capacitor is empty, or almost empty, then we say it represents a bit value of 0. It is not allowed to have a capacitor half-filled, and the circuitry is designed to make sure this doesn't happen. By grouping together eight capacitors, each of which is either full or empty, any eight-bit number—for example, 01110011—can be represented.

The job of the computer circuitry is to empty or fill different capacitors according to a set of rules called a program. Eventually, the actions of filling and emptying of capacitors manages to carry out the desired calculation—say, adding two eight-bit numbers. In a computer, the actions are performed by miniscule components of computer circuitry called logic gates. Each logic gate is made from silicon and other elements arranged in a way either to block or to pass electric charge, depending on its electrical surroundings. Inputs to logic gates are bit values, represented by a full capacitor (a 1) or an empty capacitor (a 0). (The name 'gate' is consistent with the fact that something goes into it and something comes out.)

There are several ways to choose the set of logic gates to be used for constructing a universal computer. I focus on a particular design called {XOR, AND} logic. In this design, only two kinds of gates are needed to construct any universal computer: the AND gate and the XOR gate. Each has its own rules to follow, as illustrated in FIGURE 10.1.

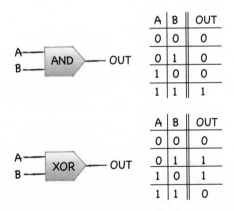

A	B	OUT
0	0	0
0	1	0
1	0	0
1	1	1

A	B	OUT
0	0	0
0	1	1
1	0	1
1	1	0

Figure 10.1 AND gate and XOR gate, each with its own set of rules for converting two inputs into one output.

The AND gate has two inputs, called A and B. Only if both inputs equal 1 does the AND gate produce an output of 1; otherwise, it produces an output of 0.

The XOR gate also has two inputs. If both inputs have the same value (00 or 11), then the XOR gate produces an output of 0. If the inputs have different values (01 or 10), then the XOR gate produces an output of 1.

An important observation is that the gate operations are not reversible. If we know the output value, we cannot deduce what the input values were that gave rise to that output.

As an example of how different combinations of gates can be cascaded to perform a specific action or operation, see FIGURE 10.2. One AND gate and two XOR gates are combined in a particular way to create the results shown in the table in FIGURE 10.2. If either of the inputs, or both, equals 1, then the output is 1; otherwise, the output is 0. This overall operation is called an OR gate, although it would better to call it an EITHER–OR gate. If you wish, you may trace through the sequential steps, checking all the values, and verify its operation.

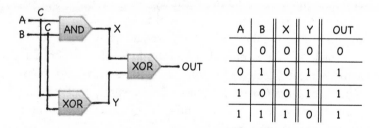

A	B	X	Y	OUT
0	0	0	0	0
0	1	0	1	1
1	0	0	1	1
1	1	1	0	1

Figure 10.2 An OR gate is constructed from one AND gate and two XOR gates.

There are three important actions of the circuit in FIGURE 10.2. First, the input bit values A and B are 'copied' at the points labeled C. The copies are passed along to the inputs of the first two gates. Copying bit values is routine in classical computer circuits, and it is accomplished by sensing the number of electrons and adding more electrons where needed. Second, the outputs of the first two gates are passed to the input of the third gate. Third, the XOR gate acts on its inputs to create an output. Notice that the number of outputs (one) is less than the number of inputs (two). This means the operation is not logically reversible; knowing the output does not allow us to deduce the two input values.

How small can a single logic gate be?

In the first all-electronic computers like the ENIAC built during the 1940s, a single logic gate was a vacuum tube similar to the amplifier tubes still used today in vintage-style electric guitar amplifiers. Each tube is at least the size of your thumb. By 1970, the revolution of microcircuitry managed to reduce the size of each gate to about one-hundredth of a millimeter. When things get much smaller than this, it's best to measure in terms of a length unit called a nanometer, which equals one millionth of a millimeter. The 1970 gate size was 10,000 nanometers. In contrast, a single silicon atom, which is the main atomic element in computer circuitry, is about 0.2 nanometer in size. By 2012, single gates in typical computers had been shrunk enough so they could be spaced apart by as little as 22 nanometers—that is, only about one hundred atoms apart. The actual working region of the gate was smaller than 2.2 nanometers, or ten atoms in thickness. This small size allows several billion memory locations and gates to be placed in an area the size of your thumbnail.

Making gate sizes much smaller than these dimensions leads to both a curse and a blessing. We leave the domain of many-atom physics and enter the domain of single-atom physics. Now the differences becomes apparent between the classical physics principles that well describe the average behavior of many atoms, and the quantum physics principles needed when dealing with single atoms. We enter the domain of random behavior, which doesn't sound good if we are trying to have a well-regulated machine do our numerical bidding.

In fact, a group of scientists led by Michelle Simmons, director of the Centre for Quantum Computation and Communication at the University of New South Wales, Australia, built a gate consisting of a single phosphorus atom embedded in a channel in a silicon crystal. This is the smallest gate that can ever be built. This gate works properly only if cooled to an extremely low temperature: −459 degrees Fahrenheit (−273 degrees Celsius). If the material is not at least that cold, the random (thermal) motion of the silicon atoms in the crystal diminishes confinement of the electron psi wave, which can leak out of the channel in which it is intended to be confined. For everyday desktop computers, which, after all, have to operate at room temperature, this leakage prevents such single-atom gates from being the basis of technology everyone can use. On the other hand, such demonstrations prove that computers can, at least in principle, be constructed at the atomic scale, where quantum physics rules.

Can we create computers that use intrinsically quantum behavior?

Given that physics determines the ultimate behavior and performance of information transfer, storage, and processing, it is natural to ask how quantum physics comes into play in information technology. Because electronic computers rely on the behavior of electrons, and communication systems rely on the behavior of photons—both elementary particles—it is not surprising that the performance of information technology is ruled ultimately by quantum physics. But there is a subtlety here. As I explained, the computer technologies that are in use today do not involve quantum superposition states to represent information. They use states that can be considered classical states of physical stuff—namely, groups of electrons.

The new question that has been asked during the past thirty-five years is: Would there be advantages to basing information technologies on intrinsically quantum mechanical states, such as superposition states and entangled states? We have already seen, in Chapter 3 that, in the context of secure communication, the answer is yes! We saw that by representing bits using the polarization of individual photons, we can design a system that enables two people to create a private key for encrypting messages in a perfectly secure manner.

The big question is: Can we create computers that use intrinsically quantum mechanical states to enhance our abilities to solve real-world problems? If such computers were ever built, they would be able to defeat certain types of data encryption methods far faster than is possible with any computer operating today. This would revolutionize the field of privacy and secrecy for computers and the Internet. An encryption key that might take thousands of years to crack using a conventional computer might take only minutes on a quantum computer.

What is a qubit?

The word *bit* is used to refer to both the abstract, disembodied mathematical concept of information and to the physical entity that embodies the information. In classical physics, it is clear that a 'physical bit' carries one 'abstract bit' of information. There is a direct one-to-one correspondence between the state of the physical bit and the value, 0 or 1, of the abstract bit.

We can also use individual quantum objects such as an electron or photon to embody a bit. In this case, the elementary physical object is called a *qubit*, short for 'quantum bit.' A qubit has two possible quantum states, such as H and V polarization for a photon, or upper path and lower path for an electron. When measured, the outcomes represent bit values of 0 or 1. But recall that we can select different schemes for performing a polarization measurement—say, H/V or D/A. Then, the outcomes may be random, with probabilities for observing the possible outcomes depending on which measurement scheme we selected. In this case there is not a one-to-one correspondence between the state of the physical qubit and the value of any hypothetical abstract bit.

The principles of quantum physics indicate great differences between the way classical bits and qubits behave. Classical bits can be copied as many time as we wish, with no degradation of information; qubits cannot be copied or cloned even once, although their state can be teleported. The state of a classical bit, 0 or 1, can be determined by making a single measurement; the quantum state of a single qubit cannot be determined by any sequence of measurements.

What physical principles set classical and quantum computers apart?

There are big differences between the kinds of gates used in classical computers and the gates that need to be used in quantum computers. Classical gates perform operations that are not reversible; knowing the output does not tell you what the inputs were. In contrast, for a quantum gate to operate properly with qubits, it must be reversible. That is, from knowing the output states you must be able to determine the input states. This requirement arises because any quantum gate operation must be a unitary process, as defined in Chapter 4.

Recall we use the word 'unitary' to refer to physical processes or behaviors that cannot be divided into individual steps each with definite, observable outcomes. Our main example was an electron (or photon) transiting from a source to a final location in a situation in which two distinct paths are possible. We emphasized that if there is no permanent trace left by the electron's passing that would indicate it took a particular identifiable path, it is wrong to say it actually took one path or the other. It's also not correct to say it took both paths. The whole process of departing one location and arriving at another must be considered as an undivided, whole process—that is, a unitary process. Such processes are reversible.

What logic gates would quantum computers use?

Since the early 1990s, scientists have been theorizing about how a universally programmable computer based on quantum superposition and entanglement could be built, and for which kinds of problems it would be best suited. Neither is an easy problem, and neither has been solved fully to date. On the other hand, great progress has been made already and prospects seem fair to good that such a computer will be a reality within, say, ten or twenty years.

A quantum computer takes in qubits as inputs, performs a sequence of gate operations on them according to a program devised by a programmer, and spits out the altered qubits as outputs. For the whole process to be unitary, the number of output qubits must equal the number of input qubits.

As there are for classical computers, there are various choices a designer can adopt for the set of gates to be used in a quantum computer. I focus here on a set that is similar to the {XOR, AND} logic described earlier for classical computers. For quantum computers, I adopt what I call {QXOR, QR} logic. Using two quantum gates, called 'quantum XOR' and 'quantum ROTATE,' a universal quantum computer can be constructed, at least in principle. What do these two kinds of gates do?

First, recall that we call the two possible states of a qubit by the names 0 and 1, and they are represented by, for example, the H-polarized and V-polarized states of a single photon. The general operating principles of a quantum computer are independent of how we choose to represent the qubits by physical objects.

The quantum XOR gate, or QXOR gate, is illustrated in FIGURE 10.3. The table of values is the same as the table shown in FIGURE 10.1. In the QXOR gate, the B qubit passes unchanged through the gate (from left to right) rather than being discarded as it is in the classical case. This makes the QXOR gate reversible and unitary. That is, the outputs are related uniquely to the inputs.

A nice way to think about the QXOR gate is to say that qubit B controls what happens to qubit A, as indicated by the arrow pointing from B to A. If qubit B is in state (0), then qubit A passes through unchanged, as in FIGURE 10.3 parts (i) and (ii). But, if B is in state (1), then qubit A's state is flipped from (0) to (1) or from (1) to (0), as in FIGURE 10.3 parts (iii) and (iv).

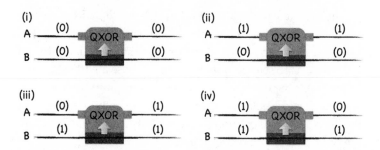

Figure 10.3 (i–iv) Quantum XOR gate. Qubit B controls the operation altering qubit A. Qubits move from left to right.

Up to this point, I haven't shown any real difference in behavior between the QXOR gate and the classical XOR gate. Now comes the punch line: What happens if the 'control' qubit B is prepared in a state that is a superposition of the (0) and (1) states? FIGURE 10.4 shows the case in which qubit B enters with the state (0) + (1) and qubit A enters as state (1). Both of the possibilities, (0) and (1), for qubit B affect the state of qubit A. The (0) possibility of the B qubit state leaves the A qubit unchanged, whereas the (1) possibility of the B qubit state causes a flip of the A qubit state from (1) to (0). The output state is therefore (1)&(0) + (0)&(1). This in an entangled state of the two qubits. In English, it reads, {qubit A is in state (1) and qubit B is in state (0)} in superposition with {qubit A is in state (0) and qubit B is in state (1)}.

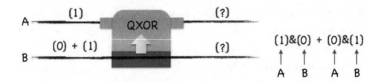

Figure 10.4 The quantum XOR gate with a superposition state at the B input causes entanglement at the output.

The state created by the QXOR gate in FIGURE 10.4 is a kind of Bell State, familiar from Chapter 9. In fact, the QXOR gate acts similarly to the Bell State generator I illustrated in FIGURE 9.3. To see the similarity, let's choose a particular way to represent the qubit states. Let's say our qubits are photons, and the vertical polarization (↑) represents qubit state (1) and the horizontal polarization (→) represents qubit state (0). Then, the output state of the QXOR gate is

$$(\uparrow)\&(\rightarrow) + (\rightarrow)\&(\uparrow).$$

This state is related to the entangled Bell State we discussed in Chapter 9, which was (↑)&(→) + (←)&(↑).

This example indicates a close connection between quantum computing gates and the devices used to create and verify Bell States. Furthermore, recall from Chapter 8 that Bell States are essential for performing experiments to test the Bell Relation, and that the observed violation of the Bell Relation proves that Local Realism is not tenable as a worldview. Quantum computing is enabled by the same quantum correlations as those responsible for the experimental invalidation of Local Realism.

The second type of logic gate needed is the quantum ROTATE gate, or QR gate, illustrated in FIGURE 10.5. This gate has one input and one output, and is reversible and unitary. If the input state is (0), then the output state is an equal superposition of the (0) and (1) states, with a state arrow pointing in the diagonal direction. If the input state is (1), then the output state is again an equal superposition of the (0) and (1) states, but with a state arrow pointing in the antidiagonal direction. Both of the arrow components 'a' and 'b' have value 0.707 (that is, the square root of one-half). According to Born's Rule, this means the probabilities to observe either outcome, (0) or (1), on making a measurement each equal 0.5, or fifty percent.

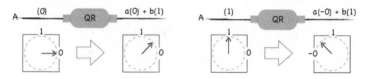

Figure 10.5 Quantum-ROTATE gate creates a superposition of (0) and (1) qubit states. The arrow-rotation diagrams are analogous to those for polarization state arrows.

These two possible output states are analogous to diagonal and antidiagonal polarization. In the case that the qubits are being represented by a photon's polarization, the QR gate is implemented easily using a special crystal that rotates the polarization state arrow by forty-five degrees in the counterclockwise direction.

Quantum computing theorists have proved that by combining sequences of QXOR gates and QR gates, any computation using qubits can be accomplished.

How would quantum computers operate?

A *classical* computer works by specifying input data, in the form of a collection of bits, and sending those bits into the processor where gates act sequentially according to a program, then reading the values of the bits at the output. A schematic example is shown in FIGURE 10.6. The input data are loaded into memory locations and the data are passed into a processor that consists of many gates. The example in FIGURE 10.6 shows an adding circuit that, through the actions of many gate, creates an output stored in designated memory locations, where they can be read. In the example, the number 5 (in binary 0101) and the number 2 (in binary 0010) are passed into the processor. The output is the binary number 0111, which corresponds to the number 7. The circuit has carried out the computation 5 + 2 = 7. The details of the gates configuration are not important for our discussion.

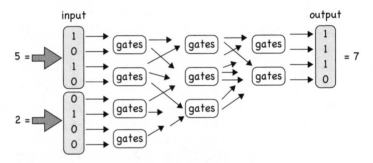

Figure 10.6 Adding circuit carrying out the computation 5 + 2 = 7.

A *quantum* computer works by specifying input data in the form of a collection of qubits each with its quantum state specified, sending those qubits into a quantum processor where gates act on them according to a program, then measuring the qubits at the output. The huge difference between the classical case and the quantum case is that only in the quantum case can superposition and entanglement exist. These quantum states can exist all the way through the circuit only if the overall action of the gates together is a unitary process. The process is unitary only if there is no way, even in principle, that a person could know any of the individual qubit values (0 or 1) in the inner part of the circuit. In the language of Chapter 4, where we discussed processes that may occur by different possible paths, we say that no qubit can be measured or leave a permanent trace of its bit value as it passes through the gates. Recall, the quantum possibilities represented by superposition states do not correspond to actual bit values or outcomes, but only to possible outcomes in the quantum physics sense. For the overall process to be unitary, the number of outputs must be the same as the number of inputs, as illustrated in FIGURE 10.7. This also makes it possible for the process to be reversible.

The beauty of a quantum computer is that it can accept superposition states at the input. Therefore, we can do a more clever thing than load two particular numbers, such as 5 and 2, at the input. We can, in effect, load all possible numbers in the range 0 (in binary 0000) to 15 (in binary 1111) at the input simultaneously! FIGURE 10.7 shows how this is done by preparing every input qubit in the state $a(0) + b(1)$. We prepare the qubits so that both of the state components 'a' and 'b' equal 0.707. According to our discussion of Born's Rule in Chapter 2, this means that if we were to make a measurement, the probabilities for either outcome, (0) or (1), would have a value 0.5 (because $0.707^2 = 0.5$). But we don't want to make any measurements at this input; that would make the overall process no longer unitary. We wait until the qubits emerge at the output of the gate circuit. There, the qubits are described by one big entangled state, which contains information about the additions of every combination of two numbers in the range 0 to 15.

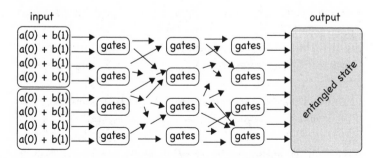

Figure 10.7 Quantum adding circuit with input qubits prepared in superposition states.

But there is a complication. When we measure the qubits at the output, we will observe random values, consistent with Born's Rule for the entangled state components. We will not learn the sum values of all the possible additions present in the computation. Evidently, to take advantage of the entanglement present in the computation, further cleverness is required.

The cleverest scientist working on this problem in 1994 was Peter Shor, who devised a method to program a quantum computer (although still a hypothetical, mythical machine) to carry out a type of calculation that is of great importance: the factoring of numbers into products of smaller numbers. I explore this topic in the next sections.

Why is factoring numbers difficult?

Quick! What two numbers when multiplied together give the result 15? If you thought of the answer, you just solved a problem called factoring. Quick! What two numbers when multiplied together give the result 35? Those two examples were easy, right?

Now consider: A prime number is a positive whole number that cannot be written as a product of other whole numbers (other than 1 and itself). Quick! What three prime numbers when multiplied together give the result 105? It would take you some time and trial and error to answer this question using only pencil and paper. (The answer is $3 \times 5 \times 7$.) Here is a really hard one: What prime numbers when multiplied together yield 435326987788326224726361413473201911779074265430773? My laptop computer has a program that finds prime factors. It found the prime factors of this fifty-one-digit number in less than one second and yielded the answer:

4788445733 × 548746972649 × 87994023186917 × 1882763238063157.

I used my laptop to try out thousands of cases of factoring numbers of different lengths. I found that, on average, the time taken to factor a number that contains N digits equals one millisecond multiplied by the quantity $2^{N/3}$. (A millisecond is one-thousandth of a second.) For example, a twelve-digit number takes, on average, 2^4 or 16 milliseconds, not much time at all. But here is the rub: The time needed to factor a number increases exponentially as the length of the number increases. For every increase in length by three digits, it takes twice as long to find the factors. The sequence of time increases is 1, 2, 4, 8, 16, 32, 64, 128, 256, 512, and so on.

This exponentially growing sequence reminds me of a parable from ancient India: The King challenged a sage to a game of chess, and if the sage won, he could name his reward. The sage said he would like only some rice grains: one grain on the first chessboard square, twice that on the second square, and twice the preceding number on each succeeding square. The King said (the equivalent of), "No problem!" But then he found that the exponential growth of the number of grains would lead to an astronomical number on the sixty-fourth square, because 2^{64} equals 18,000,000,000,000,000,000 grains of rice, which equals about 210 billion tons. The King was not happy.

For a sixty-digit number, the exponential quantity $2^{N/3}$ equals 2^{20}, which is about one million. Multiplying this quantity by one-thousandth of a second shows that it takes, on average, one thousand seconds to factor a sixty-digit number. This is still not so bad, because a computer can carry out each step of the computation so quickly. Nevertheless, the exponential growth will always catch up and overwhelm any computer. Using the program in my laptop, a 171-digit number would take, on average, 10^{14} seconds to finds its prime factors. That equals the age of the universe! Here is an example of such a 171-digit number that the program in my computer can likely never factor:

2011941382682806849295925941302824168624073409763
2982855970484357884323935164259166946211341313731
549395688627059340317115653402944300460905 6890183
2899727285719278303561 80.

This is not to say that no computer program can factor this number in a reasonable time. In fact, in 2009 a team of computer scientists factored a 232-digit number using an optimized program running on two hundred computers for two years. The computer program they used is clever enough to suppress the rapid increase of time required to factor numbers as their length increases. Nevertheless, the increase of time required is still nearly exponential, and still grows quickly beyond any human or computer capability. Using the most advanced methods known, a five-hundred-digit number would still take the age of the universe to factor.

It's amusing to me that I could create a five-hundred-digit number simply by multiplying together any two 250-digit prime numbers. As long as I keep my method for creating this number secret, then there is no supercomputer in the world that can ever discover them.

The reason factoring is difficult—in fact, 'exponentially difficult'—is that the search for the prime factors involves searching through an exponentially large set of possibilities. The search involves dividing the original number by every prime number small enough to be a candidate as a factor. If this division yields a whole number, then that number has to be tested to see if it is prime, by dividing it by every prime number small enough to be a candidate for one of its factors. Most of the possibilities tested are blind alleys and the search has to back up and search down yet more alleys. This way of searching can be compared with a crawling insect looking for a treat that sits at the very tip of one of the highest branches in a tree with a huge number of branches. Each time a wrong branch is explored, the insect needs to climb back down and start up another branch.

This situation is more than a mere curiosity. Mathematical problems that are easy to construct in one direction but nearly impossible to solve in the opposite direction are the basis for modern encryption methods. Every time you enter your credit card number to buy something from a secure website (with an address beginning with https), your computer engages with the website's computer to construct such a difficult-to-solve problem. Any hacker will have an extremely challenging, if not impossible, task of trying to reverse engineer the problem and thereby discover your credit card number. National security agencies worldwide are very interested in understanding the limits to secure encryption using such mathematical methods. They are also very interested in learning how to crack such encryption methods.

How could quantum computers solve the factoring problem?

A quantum computer, if constructed successfully, would be capable of solving certain computational problems much more efficiently, and therefore quicker, than any standard digital computer. Quantum theory, as proved by Peter Shor in 1994, predicts that a quantum computer could find the secret prime factors of a five-hundred-digit number in a reasonable amount of time—a short enough time that national security agencies around the world would love to have such a computer at their disposal.

Although the factoring problem is a rather special application of a quantum computer, it is a good example for understanding how quantum physics gives rise to acceleration of computations. Although the quantum factoring method is too complicated to explain in detail here, the basic idea is that the number to be factored—say, the five-hundred-digit number—is represented in binary form and loaded into the input, as in FIGURE 10.7. To represent any five-hundred-digit number in binary form requires around 1600 bits, so the input data will be a sequence of 1600 zeros and ones. For a quantum computer, these zeros and ones are represented by the quantum states of the qubits—one for each bit. Imagine sixteen hundred photons, each with a specific polarization state, H or V, representing the five-hundred-bit number you wish to factor. These photons pass into the quantum processor, which carries out the program devised by Peter Shor, called Shor's algorithm. Then, at the output, you measure the polarization of every photon, which yields either zero or one.

According to Shor's algorithm, the measured values tell us how to construct a set of numbers, which may or may not be the true prime factors of the five-hundred-digit number of interest. The reason Shor cannot guarantee these numbers are the true prime factors is that, as we know, the measurement of quantum objects yields results for which we can predict only probabilities rather than actual outcomes. So even the quantum computer cannot solve directly the difficult backward problem of factoring.

But now, Shor is saved by the easiness of the forward problem: When the quantum computer spits out a list of candidate factors, it is quick and easy to check to see if each is, indeed, a factor. Divide the original number by the candidate factor to see if that results in a whole number. If all the candidates are found to be factors, then we are happy we have solved a problem that would take the lifetime of the universe to solve on a classical computer. If not, then we try the whole process again, and keep trying until we succeed in finding all the true prime factors.

Shor proved that the number of tries the quantum computer requires, on average, to find the factors of a number containing a certain number, N, of digits grows much more slowly as N increases than if the growth were exponential. (With every doubling of the number of digits, the quantum calculation becomes only four times as long.)

Returning to the analogy of the insect climbing up a 'tree' of possible prime factors, Shor's algorithm allows the insect to test an entire clump of branches all at once, instead of twig by twig.

What other computer science problems could quantum computers solve?

Finding prime factors of large numbers is an interesting example of a problem that could be solved efficiently by a quantum computer, but it is not of much interest in practice for the average person or even the average scientist. What problems of greater usefulness could quantum computers solve?

Rapidly searching a large database would be enormously useful. Every time you do an Internet search for some useful tidbit of information, a complex computer algorithm is carried out in a massive computer in a massive data farm owned by one of the technological giants of industry. There are huge financial payoffs for companies that can perform data searches more efficiently. The problem gets more difficult every year as the total amount of information stored in computers worldwide continues to grow explosively.

Enter the quantum computer. Searching through an exponentially growing 'tree' of possibilities is what quantum computers could do well. Computer scientist Lov Grover discovered a quantum algorithm that can speed up database searches, although the acceleration is not as great as Shor's for factoring.

The 'bad news' is that, as far as we know, most problems of interest to mathematicians and computer scientists cannot be solved any faster using a quantum computer than a classical computer. Scott Aaronson, a computer theorist, pointed out this is because most mathematics problems don't correspond to the mathematics of quantum theory in any direct way. Examples of problems that probably would not be solved efficiently by quantum computers are playing chess, scheduling airline flights, or proving mathematical theorems automatically. To put it more precisely, not all problems with a 'time to solve' that grows exponentially with the size of the input can be solved efficiently by a quantum computer. The factoring problem is an exception only because the mathematics needed to solve it corresponds to an analogous problem in quantum theory, and because it is easy to check in the forward direction whether a candidate solution the computer spits out at random is actually a correct solution. As Aaronson says, "Admittedly, the [known] limitations do not rule out the possibility that efficient quantum algorithms for [all of the most difficult problems] are waiting to be discovered. . . . In the meantime, we know not to expect magic from quantum computers."[2]

Although quantum computers cannot magically solve all computer science and mathematics problems, they could work wonders of a nearly magical kind in the arena of basic physics and chemistry research. The next section explores their great potential in these areas.

Which physics and chemistry problems could quantum computers solve?

The history of quantum computing did not begin in computer science, but in physics. In 1981, Richard Feynman, one of the most inventive of theoretical physicists, pointed out that the basic equation of quantum theory—Schrödinger's equation—which you met in Chapter 6, cannot be solved efficiently using ordinary computers. Schrödinger's equation plays a role in quantum theory kind of like Newton's laws of motion play in classical physics theory. The difference is that, although Newton's laws describe how classical objects behave in terms of definite and perfectly predictable outcomes, Schrödinger's equation describes how quantum states change in time. Again, recall that quantum states are not in one-to-one correspondence with measurement outcomes, but represent only possibilities for outcomes.

The fact that Schrödinger's equation cannot be solved efficiently using ordinary computers is a major problem for the advancement of science. We have a fundamental equation we need to solve for predicting the probabilities of experimental outcomes; but, for sufficiently complicated situations involving many quantum objects, we can't solve it! We simply don't know precisely what the theory predicts, so we can't use it fully to advance science, engineering, and medical research. We can't design better drugs based on quantum theory because solving Schrödinger's equation for large molecules is not possible. Of course, scientists have many ways of finding approximate solutions to Schrödinger's equation, which is very helpful, but we don't have exact solutions, which might contain welcome surprises.

Feynman hypothesized that a new type of computer, which he called a quantum computer, could solve Schrödinger's equation efficiently. Since Feynman pointed this out, a lot of work has gone into trying to construct such a computer. Such a computer would itself operate according to quantum principles rather than classical physics principles, as ordinary computers do. In contrast to most computer science problems and math problems, problems that involve solving Schrödinger's equation can be turned easily into algorithms that can be carried out on a quantum computer. That is because Schrödinger's equation is the basic equation of quantum theory!

For example, in Chapter 11, I explain that Schrödinger's equation allows us to calculate the energy and the shape of the psi wave for each possible quantum state for an electron within an atom. Molecules such as the all-important DNA molecule are made of atoms arranged in ways that create their structure and allow them to perform their functions, such as encoding and propagating the genetics of a person. Because DNA molecules contain so many quantum particles—electrons, protons, and neutrons—using classical computers it is impossible to solve Schrödinger's equation exactly to understand and predict their structure and function.

To see why it is so hard to solve Schrödinger's equation for a DNA molecule using a classical computer, consider a simple example. Let's say a molecule contains a total of five hundred electrons. This is actually a fairly small molecule compared with DNA. To represent the quantum state of these five hundred electrons using the state of the bits of a computer requires representing all possible entangled states of the five hundred electrons. Each of these possible states represents a quantum possibility for a particular combination of outcomes that might be observed if measurements were made on all electrons. To keep it simple, let's say each electron could be in one of two states, labeled 0 or 1—that is, it can be thought of as a qubit. If there are two electrons, there are four possible combinations: 00, 01, 10, and 11. If there are three electrons, there are eight possible combinations: 000, 001, 010, 011, 100, 101, 110, and 111. For five hundred electrons, there are 2^{500}, or about 10^{150}, possible combinations of states that need to be considered. This is far greater than the total number of elementary particles in the whole universe! Each combination needs to be represented by a number in the computer's memory, but it is impossible to store all these numbers in any computer smaller than the whole universe. A workaround might be to break all the combinations into smaller groups and process each group separately by moving numbers into and out of the computer's memory. But the time needed to do all this moving would likely take longer than the lifetime of the universe.

This example illustrates Feynman's main point: As the size of the quantum problem to be solved gets bigger, the size of the computer needed to solve it grows even faster—in fact, exponentially. To overcome this limitation, as quantum theorist Steven Flammia has said, "We can fight quantum with quantum!"[3] We should use one collection of quantum objects, configured as a specialized computer, to imitate or 'simulate' another quantum system of interest. For example, use a quantum computer with five hundred precisely controllable qubits and a sufficient number of gates to implement a quantum algorithm to simulate the behavior of five hundred electrons in a particular molecule.

For example, let's say a chemist wants to design an improved chemical compound to improve drug activity or to increase the efficiency of solar energy generation. The strategy would be to design a quantum computer, using photons, electrons, or atoms as qubits, such that the qubits interact theoretically as they would in the actual molecules of potential interest. Then, by running different programs on the quantum computer, the structure and behavior of the candidate molecules could be simulated to find the best molecule for the task at hand.

The quantum simulation approach would be far more efficient than synthesizing chemically a vast number of candidate molecules in the laboratory and testing them out, one by one, for the best performance.

This approach to creating 'designer molecules' might make you wonder, "How would we know if the quantum computer is yielding the correct solution to Schrödinger's equation for each molecule simulated?" Such exact simulations would be impossible on a classical computer running for a reasonable amount of time. So it might seem impossible to check that the quantum computer is operating correctly. I asked this question of theoretical chemist Alán Aspuru-Guzik. His answer was, to paraphrase: "What you have to do is chemically synthesize the molecule that the quantum computer says is the best candidate, and then test it in a laboratory experiment to see if it works as predicted!" That is, forget about the pencil-and-paper mathematical calculations and even the digital computer computations. Instead, check Nature against Nature!

Why are quantum computers so hard to make?

If not sufficiently well controlled, quantum computers would be far more error prone than standard classical computers. This extreme sensitivity arises because of the difficulty of keeping all the qubits in the correct superposition state during the entire computation. Recall, for example, that the qubit superposition state (0) + (1) and the state (0) + (–1) are different states. Yes, they yield the same probabilities for a 0 or a 1 outcome if a measurement is performed, but they are quite different physically. To see this, consider a qubit represented by a single photon. The polarization state (H) + (V) is a diagonally polarized (D) state, whereas (H) + (–V) is an antidiagonally polarized (A) state. Their state arrows are perpendicular and, so, entirely different.

These states are very delicate. For example, an accidental introduction of a very small timing difference between the H and V components of the diagonal state can flip it to antidiagonal. This would disrupt completely the intended quantum computation.

Errors in the computation introduced by unwanted disturbances or 'noise' would prevent a quantum computer from yielding the correct answers if there were no way to anticipate, detect, and correct these errors as the computation proceeds. Fortunately, physicists have discovered, by using quantum theory, ways to correct such errors in a working quantum processor. The idea is to include some extra qubits at the input, whose job is to keep track of any unwanted errors. These extra qubits are entangled in a special, known way with the qubits we care about—those doing the computation. Then, by measuring the extras qubits, without disturbing the qubits we care about, we can detect an error that might have occurred. Such detection of errors is reminiscent of, but not identical to, the detection of errors in a quantum encryption key distribution setup. Recall, Alice and Bob are able to detect bit errors that were introduced by an eavesdropper trying to intercept the key information. When an error has been detected, it can be corrected before the computation proceeds.

Unfortunately, adding more and more qubits, which also need to be controlled nearly perfectly, adds greatly to the complexity of a quantum computer, making these computers very hard to build.

What are the prospects for building quantum computers?

This is a hard question to answer because the subject is a moving target. In 2017, it is safe to say there is no universal quantum computer in operation. Many small-scale demonstration projects have been carried out successfully that appear to prove that the physics on which the promise of quantum computing rests is, indeed, solid. These demonstrations seem to show that building such a computer is now 'only' a matter of ingenuity and extremely challenging engineering, rather than questions of fundamental physics.

It may take scientists ten more years or much longer to learn how to build a working quantum computer that is 'scalable.' Scalable means that if you can build a quantum processor containing, say, one hundred qubits, then it would be only twice as hard to build one with two hundred qubits, only thrice as hard to build one with three hundred qubits, and so on. That is, you don't want the difficulty of building, or the size, or amount of resources needed, to grow exponentially with the increasing size of the problem you are trying compute. This would defeat the whole purpose of building quantum computers, which is to overcome the exponential scaling problem.

What are the promising approaches to building quantum computers?

Although there are many approaches being studied, the three most promising platforms for building quantum computers are perhaps superconducting electronic circuits, isolated individual atomic ions trapped magnetically in a vacuum chamber, and isolated individual phosphorus atoms embedded in silicon crystals. Although the international race to 'get there' first is worthy of an entire book of its own, I describe only the latter approach as an example.

The research group at the University of New South Wales mentioned earlier has learned how to position individual phosphorus atoms at precise locations within a silicon crystal, the same material used for most standard computer chips. At the center of a phosphorus atom is a nucleus containing protons and neutrons. The nucleus acts like a tiny permanent bar magnet, with north and south poles. Such a magnet can be oriented with its north pole pointing up or down. A qubit is represented by the orientation of this magnet: UP represents 1 and DOWN represents 0. The orientation of the nucleus's magnet can be controlled by applying a magnetic field briefly, the force of which causes the magnet to rotate to a different orientation. This activates the quantum QR gate operation needed as part of the quantum computer operation described earlier.

To create a quantum computer, many phosphorus atoms are needed for representing many qubits, and QXOR gate operations involving pairs of qubits need to be carried out. The many phosphorus atoms are arranged in a pattern like a chessboard, with a phosphorus atom at the center of each square, which has dimensions of thirty nanometers by thirty nanometers. Recall the size of a single atom is about 0.2 nanometer, so these are very small squares!

Normally, the qubits stored in the internal magnet orientations of each phosphorus atom do not affect one another. This is 'quiet time,' when the qubit values are simply being stored. The entire silicon crystal must be cooled to an extremely low temperature: −391 degrees Fahrenheit, or −196 Celsius. This prevents the internal magnets from being buffeted by excessive jiggling of the silicon atoms making up the crystal, which could lead to the nucleus's magnets being rotated accidentally into the wrong orientations. As we discussed earlier, this would lead to errors in the states of the qubits, and require that error correction methods be used.

If researchers want to perform a QXOR gate operation between two neighboring phosphorus atoms, they activate each by passing one electron from nearby wires into each of the atoms. For reasons having to do with atomic physics, the internal magnet in each atom becomes much stronger and they begin to affect one another. (If you have ever held two magnets close to each other, you know the stronger one tends to push and rotate the other one.)

The arrangement of the two phosphorus atoms leads to the QXOR operation as follows: the magnet in one of the phosphorus atoms, called qubit B, controls what happens to the magnet in the other phosphorus atom, called qubit A. That is, if B equals 0 (magnet DOWN), then qubit A remains unchanged. However, if B equals 1 (magnet UP), then qubit A's magnet is pushed and gets flipped from 0 to 1 (DOWN to UP) or from 1 to 0 (UP to DOWN). The researchers can carry out a sequence of QXOR gate operations involving different neighboring pairs, interspersed with QR operations of any single qubit. Therefore, in principle at least, they have constructed all the needed components for a scalable quantum computer. "The great thing about this work, and architecture, is that it gives us an endpoint," says Professor Michelle Simmons, director of the University of New South Wales project. "We now know exactly what we need to do in the international race to get there."[4]

Further Reading

An introductory account of quantum computing is that by Gerard Milburn, *The Feynman Processor: Quantum Entanglement and the Computing Revolution* (Reading, MA: Perseus Books, 1998).

Notes

1 The quote is from Rolf Landauer, "The Physical Nature of Information," *Physics Letters* 217 (1996): 188–193; quote, 188.

2 The quote is from Scott Aaronson, "The Limits of Quantum Computers," *Scientific American* March (2008): 6–69; quote, 69.

3 The quote is from Steve Flammia, "Entangled LIGO," September 14, 2016, *The Quantum Pontiff*, http://dabacon.org/pontiff/?p=5188.

4 Michelle Simmons, "How to Build a Full-Scale Quantum Computer in Silicon," November 2, 2015, http://www.kurzweilai.net/how-to-build-a-full-scale-quantum-computer-in-silicon.

11

ENERGY QUANTIZATION AND ATOMS

What is energy quantization in quantum mechanics?

We now return to a more historical progression and discuss how Schrödinger's equation describes the properties and behaviors of atoms, as was Schrödinger's original motivation. This area of study is called *quantum mechanics*.

A major idea of quantum mechanics is that under certain conditions energy is quantized. 'Quantized' means discrete—that is, existing only at particular, separated values. A classical example is the discrete levels of a staircase as opposed to the continuous smoothness of a ramp. On a staircase, you can stand only at certain discrete heights. In contrast, on a ramp, you can stand at any height; in that case, the height varies continuously rather than being quantized. *Quantization* means limiting the possible values of a quantity to a discrete set of possible values. For example, replacing a ramp by a staircase would be quantization of the height. In certain situations energy is quantized.

Why is energy quantized when a particle is confined?

For a quantum particle moving freely through space, its range of possible energy values is continuous; any value is possible. Energy is not quantized. In contrast, for a quantum particle confined to a small-enough region, its energy is quantized; only certain discrete values are possible. The possible values depend on the size and shape of the region in which the particle is confined. It is found that the smaller the region, the stronger the quantization effect, in that the allowed energy values are more separated—the energy steps are larger and it takes more energy to go from one level to the next.

Why is energy quantized when a quantum particle is confined? The answer is given by Schrödinger's equation, which represents the behavior of the possibility wave, or psi wave, associated with an electron, as discussed in Chapter 6. You can understand the reason for quantization by looking at FIGURE 11.1. An electron is confined in a region bounded by the rigid 'walls,' off which the electrons bounce or reflect. Imagine the setup is in the International Space Station, so effects of gravity can be ignored for the most part. And assume there is no friction, so once set in motion, the electron would bounce forever. The electron moves freely except when it encounters a wall; therefore, all of its energy is energy of motion and there is no stored energy.

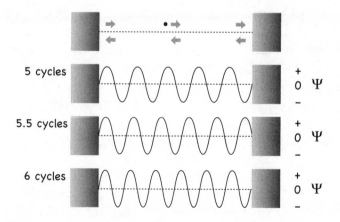

Figure 11.1 An electron transiting between two reflecting walls. In a classical picture, it follows a trajectory. In a quantum picture, it is described by a psi wave. The number of oscillation cycles of the psi wave is quantized.

In classical physics, we would use Newton's equation to predict the trajectory of the particle, which would determine its position and velocity at future times. In quantum physics, there is no concept of trajectory, only a psi-wave description. The psi wave must equal zero at the walls and beyond the walls, consistent with the fact that the probability to observe the electron at these locations is zero. To achieve this, the number of oscillations of the waves must be restricted to certain discrete values.

In the upper diagram in FIGURE 11.1, the psi wave has five complete oscillations, or full-cycle lengths. (Count them.) Because, according to Schrödinger's equation, the curviness of the psi wave designates the particle's energy of motion, five complete oscillations correspond to a particular energy. In the middle diagram, the psi wave has five and one-half oscillations, corresponding to a different possible energy. In the lowest diagram, the psi wave has six complete oscillations, giving yet another possible energy. Notice that, for example, five and one-quarter oscillations would *not* fit nicely between the walls, because the psi-wave value could not be zero at both walls, as it must be.

The psi wave will equal zero at the walls only if the number of oscillations equals an integer (1, 2, 3, and so on) or a half integer (1½, 2½, 3½, an so on). As mentioned, Schrödinger's equation tells us energy of motion is represented by the curviness of the psi wave. So, if the number of oscillations is held to certain discrete values, then the energy is also held to certain discrete values. This is energy quantization; Schrödinger's equation says, for a particle confined to a region, the particle's possible energies must be quantized.

Now we can see why energy is *not* quantized for a particle that is not confined. It can have any energy because there is no restriction on how many oscillations there could be in any particular region. So the collection of possible energies is continuous in this case.

Of course, for a particle that is described properly by classical physics, there is no direct counterpart to this behavior. A table-tennis ball bouncing back and forth between two walls can have any amount of energy in a continuous range; its energy is not quantized. Any quantum effects in such cases are covered up by the fact that the object is interacting with its surroundings (you can hear it bounce), and thus is constantly leaving behind traces of where it has been. So its motion is not a unitary quantum process and thus quantum effects are not observable.

How is the energy of an electron in an atom quantized?

The answer is again embodied in Schrödinger's equation, which asserts the following: (1) an electron in an atom moves in a way that depends on its energy of motion and its stored energy at each location, (2) the energy of motion is determined by the curviness of the psi wave representing the electron, and (3) the amount of stored energy depends on the electron's location.

Consider an electron in, say, a sodium atom (which gives the orange glow to some street lamps). An energy 'valley' is created by the attraction the electron feels toward the positively charged nucleus at the center of the atom. The electron

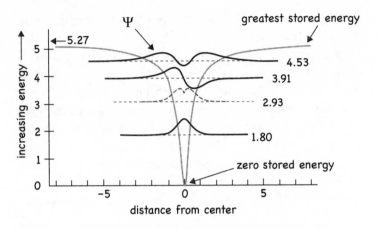

Figure 11.2 An electron in an atom moves in a stored-energy valley, the shape of which looks like the gray curve shown. The minimum stored energy is zero when the electron is at the bottom of the valley, and the maximum stored energy the electron may have without leaving the valley is 5.27 energy units. Three allowed energy values are indicated by dashed lines, with their associated psi waves. The energy 2.93, for example, is not allowed.

is pulled by electrical attraction toward the center, as if rolling down a valley side. The stored energy at each distance from the atom's center is plotted as the gray curve in FIGURE 11.2. If the electron were at the very bottom of the valley, its stored energy would be zero; it couldn't go any lower. At the other extreme, if the electron's energy of motion were to exceed 5.27 energy units,[1] it would 'fly out' and leave the valley; otherwise, it is trapped or confined within the atom's volume.

Because the electron is confined to a region, we expect the electron's energy to be quantized. In every case the psi wave must equal zero far from the atom's center, at which locations the electron has zero chance to be observed. And, in general, as the allowed energy is increased, the psi wave has more oscillations. For this shape and depth of energy valley, the three lowest possible energies are 1.80, 3.91, and 4.53 energy units, according to solutions of Schrödinger's equation. There are also higher allowed energies lying between 4.53 and 5.27 energy units.

Any energy lying between the allowed values is not allowed. The dashed curve shows an example with an energy of 2.93 units. In this case, the psi wave has a sharp kink. A sharp kink is a feature of infinitely high curviness. This is ruled out because curviness determines the energy of motion, and we know the energy of motion is not infinite. An energy value is possible only if it leads to a psi wave that is zero at the two extreme sides and doesn't have any sharp kinks or abrupt jumps in its shape.

When Schrödinger's equation is solved accurately for each particular type of atom (hydrogen, helium, nitrogen, and so on), it predicts quantized energies for each that agree in exquisite detail with experiments that measure these energies. This is why we are confident the theory is correct.

Why can't the electron come to rest at the bottom of the valley?

In classical physics, an object can certainly come to rest at the bottom of an energy valley. This might occur because friction saps the particle's energy, causing it to slow to a stop. It is known that an electron moving near a positively charged atomic nucleus loses energy by radiation; that is, it emits microwaves or light waves. In the early days of quantum physics, it was a major puzzle why an electron did not radiate away all of its energy and fall to the zero level of energy.

According to quantum physics, it's impossible for a particle's energy to be at the very bottom of the valley. Why is this? We just saw that Schrödinger's equation allows only certain possible energies, and zero energy is not one of them. An intuitive explanation for this result appeals to Heisenberg's Uncertainty Principle, discussed in Chapter 6. Recall that the Uncertainty Principle says that the more precisely you can specify the position of a particle, the less precisely you can specify its momentum, and vice versa. Therefore, if you were to specify the electron's position as being exactly at the center location—that is, at the bottom of the valley—then it could have every possible momentum. This means it would most likely have a velocity that is not zero—meaning, it is not at rest. In that case, it would move away from the center location rather quickly. On the other hand, if you were to specify its momentum as being zero, so its velocity is zero, then you couldn't also specify its location, so you couldn't say that it is located precisely at the bottom of the valley. You can't win this game!

This newfound understanding solved the puzzle that had bothered people like Niels Bohr before the advent of Schrödinger's equation. They wondered why the electron didn't fall into the atom's nucleus as a result of electromagnetic radiation carrying away all the energy, as would be expected from the models of classical physics. We now see that the answer is: "because the wavelike properties of the psi wave associated with the electron's possibilities don't allow it to be localized at one position with zero velocity." This is a drastic departure from prior classical thinking about atoms and electrons.

How does an atom absorb light?

Say the electron in a sodium atom has the lowest possible energy: 1.80 energy units. Its psi wave is shown in FIGURE 11.2 as having a single narrow peak. Born's Rule tells us that the probability to observe the electron at a given location, if we measure it, equals the square of the value of the psi wave at that location. For this lowest energy state, the most probable locations at which the electron can be found are bunched tightly near the center of the atom. The set of most probable locations is not changing or moving in time. For this reason, the state is called stationary.

The same is true of any electron state that corresponds to a particular precise energy. For any of the three allowed energy states shown in FIGURE 11.2, the set of most probable locations does not change in time. They are stationary states. In a classical way of thinking, the electron is moving around the nucleus, but in quantum physics there is no actual trajectory; all we can specify is the quantum state describing the electron, and the probabilities that state implies.

Now consider what happens if the electron is described initially by a stationary state and suddenly light is shone onto it. The electron can gain energy by absorbing some of the light. How does this happen? Let's say the electron initially has an energy of 1.80 units, so its psi wave corresponds to the one labeled 1.80 in FIGURE 11.2. Because the electron's energy is quantized, the electron cannot gain just any amount of energy. It can gain only discrete amounts that will take it from energy 1.80 units up to one of the other allowed energies: 3.91, 4.53, and so on.

Recall that light is comprised of photons, which are amounts of energy that depend on the color, or frequency, of the light. Frequency is symbolized by the letter f. Planck's Relation tells us that the energy, E, of a photon in a light beam with frequency f is given by $E = hf$, where h is Planck's constant. (In units in which time is measured in seconds and energy is measured in the units I am using here, Planck's constant has the value $h = 4.136 \times 10^{-15}$.) We can conclude that only light of very particular colors, or frequencies, can cause the electron to change its energy of 1.80 units to one of the higher allowed energies. For example, the difference in energy between 1.80 and 3.91 is 2.11 units. Planck's Relation then tells us the frequency of the light needed to cause that amount of change in the electron's energy. The frequency should be $E / h = f = 5.09 \times 10^{14}$ cycles per second. This corresponds to the frequency of the orange light emitted by sodium street lamps.

How does the electron, as represented by its psi wave, respond when light of this frequency is shone onto it? If the electron is initially in the lowest energy state, then as the light shines on it, the electron enters into a quantum superposition state with two different energies—1.80 and 3.91—as possibilities. The more intense the light, and the longer the light shines on the atom, the greater the probability that the electron could be found in the higher energy state. Because energy is conserved, we know that if the electron has gained energy, the light has lost the same amount of energy. This process is called light absorption.

An intriguing aspect of this superposition state involving two different energies is that the region where the electron is most likely to be found is no longer stationary. It moves and oscillates. FIGURE 11.3 shows frames from an animation of this moving superposition psi wave. The plotted curves show the probabilities, which equal the square of the psi wave at each location. At the initial time (*t* equals 0), the superposition of the two psi waves creates an interference pattern with a probability maximum located to the left of the atom's center, indicated by the tick mark. As time increases, the probability maximum moves rightward, passing through the atom's center and to the far right side at the time 4/8—that is, one-half

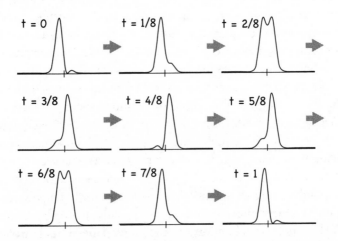

Figure 11.3 Here the electron state is a superposition of two states of different energy: 1.80 and 3.91 energy units. The psi wave, with squared values that give the probabilities of finding the electron located at various places in the atom, oscillates back and forth in time.

of a full oscillation cycle. At time equaling one full cycle (t equals 1), the probability maximum is back where it started. The full-cycle time of this oscillation equals 1 divided by the frequency, or $1/f$, which equals about 2 femtoseconds. This is an extremely short time; 1 femtosecond is just 10^{-15} second.

The frequency of this oscillation depends on the difference of the energies of the two states in the superposition. This energy difference defines a 'resonance frequency.' If the frequency of the light does not equal this resonance frequency, the light will not be absorbed and no energy will be imparted to the electron. This 'resonance phenomenon' is analogous to the classical physics situation of pushing a child on a swing. To create a large amplitude of oscillation, you must push with just the right frequency.

How does an atom emit light?

As mentioned earlier, an electron moving near a positively charged atomic nucleus loses energy by radiation; that is, it emits radio or light waves. This fact is true in both the classical and quantum descriptions of electrons.

After light has acted on the electron in the atom, the psi wave develops the oscillating behavior shown in FIGURE 11.3, indicating that the most probable location of the electron is oscillating. Therefore, the atom has a possibility of emitting light and thereby losing energy. There is a probability that a photon is created, which would travel away from the atom at the speed of light, and the electron would return to the state of lowest possible energy: 1.80.

Another type of emission process can occur if the electron is initially in the stationary state labeled 3.91 in FIGURE 11.2. In this case, the electron probability is not oscillating back and forth as in the previous example. Einstein pointed out that the electron can still emit a photon and drop to the lower energy state of the atom. He called this 'spontaneous emission.' Such an emission process is considered to be a quantum mechanical effect, consistent with Heisenberg's Uncertainty Principle.

What has become of the classical physics idea that an electron in an atom orbits around the nucleus?

The stored-energy valley I showed in FIGURE 11.2 is actually a one-dimensional slice through a three-dimensional surface. It represents motion of the electron on a line passing nearly through the atom's nucleus. To get a more complete picture, I show the same surface in two dimensions in FIGURE 11.4. It looks like a funnel. From this picture you can visualize that, in a classical description, an electron could move back and forth along a line through the center of the funnel or it could orbit in a circular path around the center of the funnel.

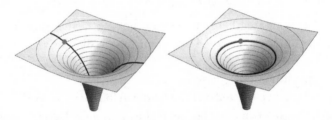

Figure 11.4 The stored-energy surface for an electron in an atom, plotted in two dimensions. Classical trajectories are shown as the bold lines through the center or orbiting around the center.

In the case of a quantum particle, how can we visualize this quantum concept of circular orbiting motion? It should appear as a wave circling the center of the funnel. FIGURE 11.5 shows two sketches of possible circling waves, one with four complete wave oscillations around the circle, and the other with ten. The wave with more oscillations has the greater curviness, and therefore corresponds to the greater energy. Also shown are examples of two impossible waves, which attempted to

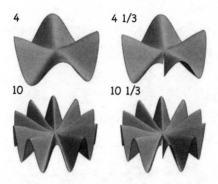

Figure 11.5 Circular waves representing an electron's psi wave in the stored-energy surface shown in FIGURE 11.4.

have either four-and-one-third or ten-and-one-third complete oscillations around the circle. These failed waves have sharp jumps and kinks, and therefore cannot exist, because they are inconsistent with Schrödinger's equation for a particle having finite energy.

These circular waves are stationary, as are the waves in FIGURE 11.2. The probabilities of finding the electron at various locations do not change in time. (The stationary waves correspond to the interference pattern of two waves circling around the center in opposite directions.) Being stationary is a characteristic of states having definite energy.

The exact values of the possible energies of an electron in an atom depend on the two kinds of motion shown in FIGURES 11.2 and 11.5: motion through the center of the funnel-shaped energy surface and motion circling around the center. The details are determined by Schrödinger's equation.

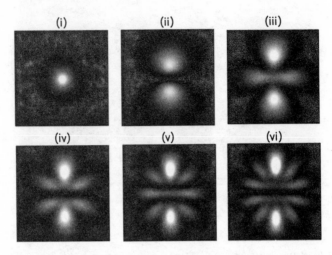

Figure 11.6 Two-dimensional slices of electron psi waves for an electron in an atom, with energy increasing in images (i) through (vi). (Created using *Atom in a Box*, http://daugerresearch.com/orbitals. With permission of Dauger Research, Inc.)

What do electron psi waves look like in three dimensions?

Real atoms are three-dimensional, of course, and the illustrations I showed earlier are crude representations, but still retain key features of the physics. In FIGURE 11.6, I show some examples of stationary psi waves from exact solutions of Schrödinger's equation for an electron in a hydrogen atom. The images are two-dimensional slices through the three-dimensional probability plots. You can see that as the energy increases, the psi wave becomes more curvy—that is, it has more rapidly varying spatial structure, indicating greater energy of motion.

Note

1   The energy units in atomic physics are called electron volts, but we won't use such jargon.

12

APPLICATION: SENSING TIME, MOTION, AND GRAVITY WITH QUANTUM TECHNOLOGY

What are quantum physics–based sensing technologies?

Quantum physics–based sensing technologies are those that rely on quantum physics principles for their working. If classical physics were the whole story, we would not have access to these new technologies. In Chapter 5, I introduced the technology of sensing and introduced a sophisticated but nonportable technology for sensing gravity. In this chapter, I discuss how the quantization of energy in atomic states, as explored in Chapter 11, leads to atomic clocks, atomic accelerometers, and atomic gravity sensors. These are constructed to sense time, acceleration, and the strength of gravity, respectively. Let's first discuss sensing time.

What is a scientific definition of time?

Isaac Newton, the greatest physicist of the seventeenth century, reckoned that time was like a steadily flowing stream that sweeps everything along from the past to the present. He wrote, "Time, of itself, and from its own nature, flows equably without relation to anything external."[1] The modern idea of time is summarized by physicist David Mermin, who wrote:

> While it is commonly believed that there is something called time that is measured by clocks, one of the great

lessons of [Einstein's theory of] relativity is that the concept of time is nothing more than a convenient device for summarizing compactly all the relationships holding between different clocks.[2]

This might seem a strange way to define time, but it rests on an underlying idea that the universe is orderly and that events occur in a synchronized way. Yet it denies there is a 'master clock' somewhere that somehow directs this synchronization.

We can use Mermin's description of time as a pragmatic viewpoint that helps us understand how to build and test highly accurate clocks. Let's say we own a collection of pretty good, fairly expensive clocks, which we always keep in good running condition. One day, we suspect one of the clocks is operating improperly. How would we test this? The obvious way is to compare that clock against all the other nearby clocks. (These clocks should not be moving relative to the clock to which they are being compared, so that effects of relativity don't cause discrepancies.) If all of those clocks are very close in their indicated time, and the suspect clock disagrees with all the others, we would be justified in believing the error is in the suspect clock and not in all the others.

What is a clock?

A good clock is any object or device that performs regular, identical motions repeatedly and counts them. As I said earlier, we can test whether a clock is 'good' only by comparing it with a collection of other clocks that we believe to be as good or better. The clocks should all have the same rate, or frequency, of ticking. Furthermore, to observe more easily any discrepancies between two clocks, they should be ticking at the highest frequency possible. This allows us to note any discrepancy sooner rather than later, because one clock will more quickly get ahead of the other.

Examples of things that make pretty good clocks include a weight oscillating on a spring, a vibrating quartz crystal (as in many wristwatches), and a swinging pendulum (as in grandfather or longcase clocks). A pendulum is a weight (called a bob) that hangs on a fixed-length rod. A pendulum retains the same timing of its swing even as it gradually 'runs down' in the distance of its swings. That is, as it runs down, its frequency remains constant. The length of rod that suspends the bob sets the time interval for it to undergo one complete swing cycle. This time is the 'full-cycle time.' As long as the rod length doesn't change, the rate of swinging is constant. To keep the clock from running down and stopping, friction is kept to a minimum, and small amounts of energy in the form of 'pushes' are added periodically. Pendulum clocks of this type can operate with a precision of about one second in twelve years.

Any naturally repetitive motion can be used to make a clock. Mechanical oscillators have the nice property that they naturally oscillate at a particular frequency, which depends on their properties and construction. We call this particular frequency the 'natural resonance frequency' of the oscillator.

How can we make clocks identical?

For a set of clocks to stay in time with each other, each clock in the collection should be as nearly identical to the others as possible. Here, quantum physics enters the story in a surprising way.

A fundamental fact of Nature discovered through the study of quantum physics is that all elementary particles of a given type are exactly identical. That is, all electrons are identical, all photons are identical, all neutrons are identical, and so on. The identicalness of like elementary particles has been accepted as a basic fact; it cannot be deduced from other known facts.

Therefore, I list it as one of the Guiding Principles of quantum physics:

> Guiding Principle #7—All elementary particles of a given kind are identical.

This doesn't mean, for example, that all electrons need to have the same energy, any more than two identical automobiles (if such existed) would always have the same speed. Yet they are still identical. The same applies to photons of light. A red photon is identical to a blue photon in composition, although the two have different energies and, therefore, colors.

If you think about it critically, you might conclude there is no reason to presume that Nature is made of any kinds of exactly identical building blocks. It could be that, at the microscopic level, there is just a swirling 'swamp' of infinite variety, with no exactly repeated objects. Quantum theory, on the other hand, rests on the identicalness of elementary particles as a basic hypothesis, and the theory has been validated by comparing the outcomes of many careful experiments with the theory's predictions.

Why do elementary quantum objects make the most perfect clocks?

By now, you can probably answer this question. The best clocks are those timekeepers of which you have several identical copies. Elementary particles are identical, and each behaves as if it contains an internal clock. The same is true for two atoms of the same type; for example, two sodium atoms with the same composition of electrons, protons, and neutrons are identical. Furthermore, energy quantization of electron states in atoms ensures the clock oscillation frequency is defined sharply and is the same for any two atoms of the same kind. Therefore, isolated atoms are ideal candidates for making superior clocks! So-called quantum clocks or atomic clocks are based on the timekeeping abilities of single atoms or a small collection of atoms.

Why are good clocks technologically important?

The global positioning system, or GPS, is something many of us use nearly every day, and its operation depends critically on having excellent clocks installed on satellites that broadcast GPS radio signals to our mobile positioning devices. Every time you use a GPS to navigate to a new location, you are using atomic clocks on at least four satellites.

Another use of good clocks is for synchronizing computers and other devices connected to the Internet. Messages and other data travel from computer to computer using either brief electrical pulses or laser light pulses moving in fiber-optic cables. In the near future, many applications will require highly precise timing across the Internet. Examples include remotely controlled surgical procedures (telesurgery), driver-less cars, operation of the electrical power grid, and financial transactions.

Also, exquisitely good clocks have many uses in scientific research. In one of the most important examples, in 1971 Joseph Hafele and Richard Keating synchronized two atomic clocks, then put one into a jet airplane, which flew westward once around Earth. When the clock that had flown was compared with the clock that stayed on the ground, the flown clock was found to be 275 nanoseconds younger! That is, it had ticked fewer times. This result agrees nicely with the predictions of Einstein's theory of relativity, which deals with the relation between time and space. This kind of experiment has been repeated since then, using better and better clocks, always in more-perfect agreement with relativity theory.

In fact, this slight slowing of the flown clock has to be accounted for in the operation of the GPS system; otherwise, timing errors would have you driving in the wrong direction! So if anyone tells you they don't believe in relativity theory (because it seems weird or something), ask them if they use a GPS.[3]

How precise are today's atomic clocks?

Einstein's general theory of relativity predicts that all clocks, including quantum ones, run slower when placed in a region of stronger gravity. In fact, very near a black hole, where gravity is as strong as it can possibly be, a clock's rate of ticking would slow to zero compared with a clock far from the black hole. If you could get near a black hole and could survive to tell the tale, the effect would be so strong that you could use any old wristwatch to observe gravity-caused clock slowing by comparing your watch with one worn by a friend far from the black hole. Alas, the strongest gravity we have at our disposal for such tests is that created by Earth, which is far weaker than that near a black hole. To test clock slowing by Earth's gravity, we need a very precise clock indeed.

Fortunately, atomic clocks can be made so precise that a small difference in the strength of gravity, such as that seen at 10,000 feet of elevation compared with that at Earth's surface, can be observed. In a demonstration of just how amazingly precise modern atomic clocks can be, in 2010 physicist James Chin-Wen Chou and colleagues at the National Institute of Standards and Technology, or NIST, in Boulder, Colorado, built two atomic clocks and placed one on a table and one on a lifting jack next to the first clock. They ticked in synchrony so precisely that Chou calculated that neither would gain nor lose one second in less than 3.7 billion years! Then he raised the clock on the jack vertically by a third of a meter (about one foot). Remarkably, he was able to observe that the raised clock ticked ever-so-slightly faster than the one resting on the table. For each one second of time elapsed, the faster clock was observed to gain about 10^{-17} of a second. This tiny but observable gain occurred solely because gravity is ever-so-slightly weaker the farther you move from Earth's surface.

How do basic atomic clocks work?

Atomic clocks are based on Planck's energy–time relation—namely, the fact that the ticking rate, or frequency, of a quantum object's 'internal clock' equals the object's energy divided by Planck's constant. There are various ways to take advantage of this fact to build a clock.

Recall from Chapter 11 that an electron confined in the volume of an atom can have only the allowed quantized energies that satisfy Schrödinger's equation. The quantization of energy is the key aspect of atoms that make them such good objects for building clocks. Each atom of a given type—say, sodium—has exactly the same resonance frequency, which is the frequency of light needed to cause an oscillation of the electron's psi wave. In the absence of energy quantization—a true quantum effect—we wouldn't be able to take advantage of this feature to build clocks. Without quantum physics, we wouldn't have near-perfect clocks; and without such clocks, we wouldn't be able to keep our modern technological world synchronized.

The world's time standard is currently based on the atomic cesium clock, so let's discuss how it works. Cesium is an element that has fifty-five electrons per atom surrounding its nuclei in a kind of 'cloud of possibilities,' represented by a psi wave. In its pure form, cesium is a soft, shiny metal. When heated to a high temperature, it melts, and single cesium atoms evaporate from the liquid, forming an atomic vapor above the liquid (like water vapor escaping a heated tea kettle).

The electron cloud inside each cesium atom has a natural resonance frequency at which it tends to oscillate naturally, as discussed in Chapter 11. Because of energy quantization, this frequency equals the difference of energies of the lowest energy state and the next-lowest energy state, divided by Planck's constant. If we want to get the electron cloud oscillating, we need to push it periodically at the proper frequency. In the previous chapter I made the analogy of pushing a child on a swing to achieve the largest swinging motion possible. For an electron, this 'pushing' is done using oscillating microwaves or laser light.

For a cesium clock, the resonance frequency is 9,192,631,770 pushes per second. Fortunately, this high frequency of pushing can be created using microwaves—the same as are used in a microwave oven to heat up the water molecules in your food. (The electron psi wave in a water molecule oscillates and absorbs microwave energy most strongly if pushed repetitively by oscillating microwaves with a frequency of 2,450,000,000 cycles per second, which is somewhat less than the natural resonance frequency of the cesium atoms in an atomic clock.)

Here is how the basic-model cesium atomic clock works: Microwaves are generated by an electronic device called a magnetron. An electronic circuit tunes the generated microwaves, much as a musician tunes a guitar or a violin to get it into correct pitch. When the frequency of the microwaves is tuned just right—to 9,192,631,770 cycles per second—this results in the most effective pushing by the microwaves, and the electron cloud begins oscillating maximally. By monitoring continually how much microwave power is being absorbed, and by making small tuning adjustments, the microwave frequency can be held to exactly the resonance frequency of the cesium. As soon as the frequency has been stabilized this way, a separate electronic circuit counts the oscillations of the microwaves and, after exactly 9,192,631,770 oscillations, the circuit indicates that one second has passed.

In fact, one second of time is now defined by the science community by counting 9,192,631,770 oscillations of cesium-atom electrons. This is now the internationally accepted definition of one second. Although this might seem arbitrary (it is!), this definition is convenient, because any technically knowledgeable person can build such a cesium atomic clock and use it to measure time. Such clocks are now so compact and cheap they are deployed in many locations, including on satellites and in cell phone towers.

How do the most advanced atomic clocks work?

There are more advanced models of cesium atomic clocks that keep time far more accurately than the basic, portable model I just described. These advanced models are being improved and updated continually, in a friendly timekeeping competition between the world's top frequency-standards research laboratories.

The official timekeeping standard currently in the United States (2017) is called NIST-F1, which is operated at the National Institute of Standards and Technology mentioned previously. It gains its improved timekeeping ability by keeping each cesium atom exposed to the microwaves for longer periods than is possible in the simple design I described earlier. The challenge is that cesium atoms, like juggling balls, when released, drop toward the ground under the force of gravity. When an atom drops out of the region where microwaves are present, or when it hits the floor of the metal chamber holding the atoms, it is no longer 'in the game.' Limiting the time each atom interacts with the microwaves limits the precision with which the microwave frequency can be determined.

So the goal of the game is to observe each atom for as long as possible. But hot atoms in a vapor move at thousands of meters per second, so in a metal chamber with a size of one meter you would have only about one thousandth of a second to observe an atom before it hits a chamber wall. Therefore, your first goal is to slow down the atoms. This you can do by shining laser light beams directly onto the atoms from six directions, as shown in FIGURE 12.1. The cesium atoms are placed initially at location 1, where laser light of the proper color nudges them into a ball-shaped region and slows them to an average speed of about 0.03 meter per second.

Then, the laser at the bottom sends a strong enough burst of light to launch the ball of atoms upward through the region where microwaves are strongest, called the microwave cavity. The atoms reach a peak height at location 2 (in FIGURE 12.1) and then, like the juggling balls mentioned earlier, they fall back toward Earth, again passing through the microwave cavity. When the atoms reach the light beam produced by the 'probe laser' (which is different from the other six lasers), they are either in a state that can deflect some of the probe laser light into the light detector shown at location 5, or they are not. If the microwaves are tuned properly to the cesium atom's resonance at 9,192,631,770 oscillations per second, then when they reach the probe laser beam they will be able to efficiently deflect some

(i)   (ii)

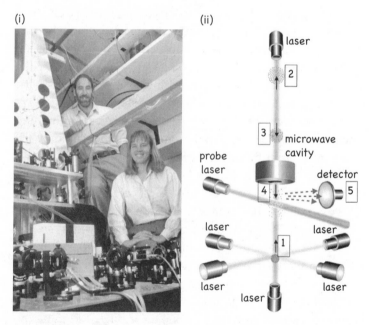

Figure 12.1 (i) Steve Jefferts and Dawn Meekhof at the National Institute of Standards and Technology with their F1 clock. (ii) The clock keeps time by launching slow cesium atoms from location 1 up to location 2, from which they fall back down. If the microwaves through which the atoms pass have just the right frequency, at location 4 the atoms deflect some light from the probe laser beam. Courtesy of the National Institute of Standards and Technology, © Geoffrey Wheeler, 1999.

of the light into the detector. If too little light is being received at the detector, the frequency of microwaves is adjusted slightly to maximize the amount of light being received. This adjustment results eventually in the microwaves oscillating at precisely the natural resonance frequency of cesium atoms—that is, at 9,192,631,770 oscillations per second.

How does the time each cesium atom stays in the NIST-F1 clock compare with the same time in the basic-model cesium clock? It takes the atoms about the same time to travel up to the peak and back down as it would take a juggling ball if you were to toss it yourself. That amount of time is about one second, which is one thousand times longer than in the basic-model cesium clock we discussed earlier. Therefore, based on the argument I made earlier, this clock should be at least one thousand times more precise than the basic model. By this method, the F1 clock keeps time to within one second in about one hundred million years.

Our next example of applying quantum physics to technology is atomic accelerometers, which also go by the general name 'inertial sensors.'

What are inertial sensors?

Inertial sensors detect small changes in the acceleration of an object or in the strength of gravity. Atomic physics–based technology is leading the way to build more and more precise inertial sensors.

The idea of inertia or momentum is that objects tend naturally to resist changing their speed and their direction of motion. Precise measurements of inertia allow physicists to test the theory of gravity—that is, Einstein's general theory of relativity—at a deep level. For example, Einstein's theory predicted the phenomenon of gravitational waves, which in 2015 were detected for the first time using advanced inertial sensors. Such tests of relativity reveal a lot about the structure of space, time, and the universe at large.

After explaining how conventional accelerometers work and how they are used, we explore how quantum physics can be used to improve their performance and thus their range of uses.

What is an accelerometer?

Analogous to a thermometer, which senses temperature, and a speedometer, which senses speed, an accelerometer is a device that senses acceleration. Acceleration means the rate of change of velocity—that is, how rapidly something's velocity is changing. For example, if you are moving at a steady speed in your car, your speed is not zero, but your acceleration is zero. If you step on the gas pedal, your acceleration becomes positive; if you hit the brake pedal, your acceleration becomes negative. An accelerometer senses and measures the rates of changes in speed and direction.

The human body is a crude accelerometer; it can feel changing speed. If you stand in an elevator with your eyes closed, you can sense when it starts accelerating—that is, when its speed starts changing. You can also sense whether it is starting to move upward (you feel heavier) or downward (you feel lighter).

In fact, according to Einstein's general theory of relativity, the effects of acceleration are indistinguishable from the effects of gravity. If you were standing in a closed elevator suspended in outer space and a very large comet were to pass right underneath, you would feel ever-so-slightly heavier as a result of the force of gravity the comet exerts on you. But you couldn't distinguish that feeling from the feeling created by someone pushing up from under your elevator and accelerating it in the 'upward' direction. In both cases, you would feel slightly heavier. If you had an accelerometer along with you in the elevator, it also could not tell the difference between a force of gravity and an acceleration applied upward to the elevator.

The good news, then, is that you can use the same device to sense either acceleration or gravity, but you have to be careful to know which one you are actually sensing. This you do by context. For example, you look out a window of the elevator to see what is going on, such as a comet passing or someone pushing upward on the elevator.

How do conventional accelerometers work?

Off-the-shelf accelerometers typically work by suspending a small weight by a stiff spring and, using electronic means, detecting the distance between the weight and a fixed surface near it. When the device is subject to a change of acceleration or a change in gravity, the small weight is pulled 'downward,' away from the fixed surface, or boosted 'upward,' away from the fixed surface. The direction it moves depends on whether the acceleration or gravity change is positive or negative. Such devices can be miniaturized to the size of a fingernail.

An even more compact design uses a special kind of crystal (piezoelectric) that is sensitive to acceleration. The acceleration causes the crystal to compress slightly, which in turn creates a voltage signal. These are small and cheap, but not very sensitive or precise.

What are accelerometers good for?

Accelerometers in smartphones allow you to give it commands simply by moving or shaking it. Accelerometers mounted in an automobile are used to sense a collision of the car with another object so the safety airbags can be deployed quickly.

By mounting an accelerometer in a vehicle and measuring the force of gravity in three perpendicular directions (forward, left, and down), you can determine whether the vehicle is titled relative to vertical. That is, you can determine the vehicle's orientation without looking out the window. By mounting several accelerometers in the vehicle, you can tell if it is rotating and, if it is, how fast. This is useful for self-driving cars or autonomous flying drones.

If the accelerometers were good enough, by recording their acceleration readings continuously, you could keep track of the direction and distance traveled from a known starting point by a vehicle containing the device. Ultimately, such methods could replace the GPS for monitoring a vehicle's location as it moves. The advantages of such a system are that it could work underground or under the ocean, where GPS satellite signals cannot penetrate; it could work well during solar storms that disrupt normal electronics on Earth; and it would be immune to electronic jamming by an adversary.

What are gravimeters and what are they used for?

Extremely sensitive accelerometers designed to measure acceleration in only the up–down direction are called gravity meters, or gravimeters. The gravimeter is held stationary so its readings reflect the strength of gravity, not any motion of the device. As discussed in Chapter 5, gravimeters are used for mapping the varying strength of gravity over a geographic area. Such mapping can detect and even image underground mineral deposits and archaeological or industrial structures.

How do conventional gravimeters work?

A sensitive method for measuring gravity is based on an idea that could have occurred to Isaac Newton when watching his proverbial apple fall from a tree. After an apple is dropped from some height, the quicker it is seen to accelerate, the stronger the force of gravity can be inferred in the immediate area. The challenge in using this 'dropping' method for measuring the strength of gravity is in observing and measuring the extremely small differences of acceleration that occur in different locations as a consequence of the differing gravity there. Excellent methods have been devised using laser light to monitor the height of a falling object, which is typically a mirror that reflects laser light as it falls. By recording continuously the height as an object falls, its acceleration can be calculated. Then, the whole device is moved and the dropping experiment is repeated. After many locations are tested, a map of the gravity strength can be made, looking much like a contour map of different elevations in a geographic area. The overall shape of the gravity map points investigators to regions likely to contain underground oil or mineral deposits.

The falling-mirror method can achieve repeatability and precision of a few parts per billion. Although this is an impressive feat of technology, even better accelerometers are needed for creating detailed images of what lies underground. Furthermore, current accelerometers are not yet good enough to allow 'absolute navigation' not relying on GPS. Current research is aimed at using quantum effects to increase the sensitivity of gravimeters (how weak an effect can be detected) and their precision (how small a difference in adjacent readings can be detected).

How does a basic quantum gravimeter work?

In Chapter 5, I described the first experiment to detect the effects of gravity in a quantum mechanical setting. That technique used neutrons in an interferometer. The interferometer was carved out of a single crystal of silicon and has three parallel silicon plates through which neutrons can pass in a straight line or in a direction deflected upward. Each neutron has two possible paths for reaching a detector, so quantum interference affects the probability of reaching the detector and registering an event. We use de Broglie's relation between a particle's momentum (speed multiplied by mass), and the marks on its 'quantum ruler,' which is expressed by

$$Distance\ between\ marks\ on\ the\ quantum\ ruler\ =\ \frac{h}{Momentum},$$

where h is Planck's constant. When the neutron comes into the interferometer with a certain speed, it has an associated quantum psi wave with a full-cycle length (distance between marks on the quantum ruler) determined by its momentum. If the neutron deflects and 'climbs' upward against the force of gravity, it slows down. This means the quantum psi wave has a slightly longer full-cycle length in the upper path than in the lower path. Therefore, when the parts of the psi wave meet, they interfere in a way that depends on how much slower the wave travels in the upper path than it does in the lower path. Researchers found that if they tilted the whole apparatus gradually, so the neutron had to climb higher and higher in the upper path, then the neutron beam was observed to switch back and forth between two detectors. How rapidly this switching back and forth occurred was seen to change as the apparatus was tilted gradually, and this dependence indicates the strength of gravity in the vicinity of the apparatus. Unfortunately, the neutron interferometer is not portable and so is not a practical gravimeter.

How do advanced quantum gravimeters work?

Present-day quantum gravimeters are based on atoms controlled by lasers, in an apparatus similar to that used for the atomic clock described earlier. In 1991, such a quantum gravimeter was demonstrated by physicists Mark Kasevich and

Steven Chu (who later was awarded the Nobel Prize and also served as US Secretary of Energy). This gravimeter is based on quantum interference of falling atoms, rather than monitoring a falling object such as a mirror. In 1999, the device achieved a precision of a few parts per billion, and, fifteen years later, newer versions constructed in several laboratories around the world have achieved precision of a few parts per ten billion. Intensive efforts are underway to create portable, commercially available gravimeters with such high performance.

The first stage in the operation of Kasevich and Chu's atomic gravimeter is shown in FIGURE 12.2. Inside a high-vacuum chamber, about one million cesium atoms are collected into a small cloud, shown as the gray-shaded circular area at the lower left. Every atom in this cloud has its internal electron in a particular quantum state, represented by the psi wave labeled 'a.' This is the circular psi wave corresponding to one of the possible energies the electron can have inside the atom, as discussed at the end of Chapter 11.

As FIGURE 12.2 shows, the cloud of atoms is launched, using laser techniques, in an upward–diagonal direction. Then, a dim laser pulse is flashed onto the atoms from below, labeled 'kick 1' in the figure. From now on, let's talk about just one atom because each exhibits the same behavior in this experiment. The electron in the atom may or may not acquire energy from the laser light; the brightness of the laser is adjusted so the electron has a fifty-percent probability of doing so, in which case its quantum state changes to the state represented by the psi wave labeled 'b' in the figure. If the electron's quantum state changes, the electron gains energy, and at the same time the laser light loses energy. Because the light loses

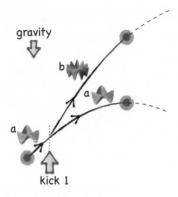

Figure 12.2 When a moving cloud of atoms is 'kicked' by a pulse of laser light, two quantum possibilities are created: it may or may not gain energy and momentum from the light, resulting in two possible quantum paths for each atom.

energy, it also loses momentum. At the same time, the atom gains momentum in the upward direction, so its trajectory arcs higher upward than if the atom had not gained momentum. Therefore, two quantum paths become possible for the atom, a lower one and an upper one. You might see that this begins to look like an interferometer.

Recall, as always, that a given atom does not actually take one path or the other, nor does it take both. The motion of the atom is a unitary quantum process and so it cannot be broken into observed paths. But, as in our previous discussions of interferometers, the two possibilities do exist, and path interference affects our predictions of where the atom might end up when measured at the end of the experiment.

The remaining steps in the operation of the atomic gravimeter are shown in FIGURE 12.3. One half-second after the first laser pulse is applied, two more pulses are applied, one from the top and the other from the bottom. These pulses (kick 2) are brighter than the first pulse, and result in a one-hundred-percent probability of changing the electron's quantum state

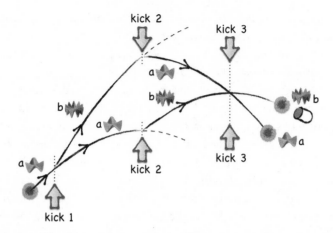

Figure 12.3 In an atomic gravimeter, atoms are launched, then kicked three times with laser pulses. The probability of the atoms arriving at detector b depends on the strength of gravity in the vicinity of the gravimeter.

in the atom. When kick 2 acts on an atom in the lower path, it changes the electron's state from 'a' to 'b' and induces an upward motion of the atom. In contrast, when kick 2 acts on an atom in the upper path, it changes the electron's state back to 'a' and causes a steeper downward motion of the atom.

After another half-second, the two possible paths come together and can interfere if two more dim laser pulses (kick 3) are applied at this time. These pulses result in a fifty-percent probability of changing the electron's quantum state and changing the atom's path.

If an atomic detector is placed in the 'b' path, as shown in the far right of FIGURE 12.3, there are two possible quantum paths an atom could be said to follow to reach this detector. It could take the upper path after kick 1, then deflect upward at kick 3. Or it could take the lower path after kick 1, then fail to deflect at kick 3, staying on its original course to the detector. These two quantum paths interfere in a way that depends on the synchronicity of the relationship between the atom's psi waves (or quantum rulers) at the meet-up location where kick 3 acts. If these two psi waves come together 'in phase,' the atom will go to the detector. If these two psi waves come together 'out of phase,' the atom will not go to the detector, but will emerge in the other outgoing path.

As with the neutron interferometer discussed earlier, the detector outcome depends on the strength of gravity in the vicinity of the experiment. If an atom climbs up against gravity along path 'b' after kick 1, it will slow, and its quantum-ruler full-cycle length will become longer in that path relative to that in the lower path. When the psi waves come together again at the meet-up location, their full-cycle lengths are again the same, but one ruler has been shifted relative to the other, affecting the outcome of the interference. Therefore, by observing the probability of the atoms arriving at the detector, as the gravimeter is moved slowly from place to place, very small changes in gravity's strength can be measured.

Can atomic interferometers detect gravitational waves?

Einstein predicted in 1916 (based on his general theory of relativity) that, during extreme situations, such as two heavy stars orbiting each other at a close distance, gravity itself may form into waves that can travel throughout the universe. Gravitational waves are similar to radio waves in that they travel at the speed of light, but their 'medium' is the gravitational field instead of the electromagnetic field.

Many experiments have been carried out in attempts to detect gravitational waves, but only in 2015 did scientists succeed in this quest. The successful method was achieved by scientists at the Laser Interferometer Gravitational Wave Observatory (LIGO). The process uses laser interferometers in which light interferes after traveling in two possible paths. There are two ways in which quantum technology can be used to improve the capabilities of gravitational wave detectors. The first is to increase their sensitivity further, which would allow astronomers to detect such waves originating from more distant sources than is possible now. The second is to help bring down the now-enormous cost of such detectors (more than $US 500,000,000 for the LIGO interferometer).

The first way to improve gravitational wave detectors is to use quantum physics techniques to reduce the amount of uncontrolled jitter in the phase (a wave's oscillation timing) of the laser light as it traverses the interferometer paths. Nergis Mavalvala, a professor at the Massachusetts Institute of Technology and one of the team leaders on this project, says it's "like trying to measure the length of a piece of paper while the ruler's tick marks keep wiggling and moving about." She explains, "Because this noise causes the tick marks on our meter stick to jitter, we want to reduce that."[4] It is possible to engineer the state of a light beam so its phase is better controlled and defined than in an ordinary laser light beam. This technique is called quantum squeezing. Although it reduces the uncertainty in the light's phase, it increases the uncertainty in its instantaneous brightness. This is another example of Heisenberg's Uncertainty Principle, which we discussed in Chapter 6. Here, instead of referring to the uncertainties in a particle's position and speed, the same principle applies to uncertainties in a light beam's phase and brightness. This quantum method for reducing the jitter in the LIGO interferometer setup will likely become activated in 2017, thereby increasing the range of the LIGO interferometer for detecting gravitational waves from more distant astronomical sources.

Given the high cost of such instruments, physicists including Mark Kasevich (mentioned earlier) have proposed that gravitational waves might be detected using atomic interferometers of the type described here, instead of the large laser interferometers used at LIGO. In this scheme, if a gravitational wave passes through an atomic interferometer, the strength of gravity would vary in time in an oscillating manner and would thus be detectable. Unfortunately, a single interferometer of this type cannot be made sensitive enough to record such a small change. The solution, according to Kasevich, is to place atomic interferometers in three satellites orbiting Earth, separated in a triangular pattern by about one-thousand kilometers. The satellites would be spaced apart far enough so that, as the gravitational wave passes, each atomic interferometer would record a slightly different strength of gravity at different times. Although still expensive, such a scheme would cost less than the LIGO interferometer, and it would have the advantage of being isolated from unwanted Earth-based vibrations, making the instrument even more sensitive, and thus more useful.

There are undoubtedly many more scientific and technological applications of atom interferometers we will see in the future.

Figure Note

The artwork illustrating the clock operation in Figure 12.1 is adapted from artwork by NIST, "NIST-F1 Cesium Fountain Atomic Clock," rev. September 21, 2016, http://www.nist. gov/pml/div688/grp50/primary-frequency-standards.cfm.

Notes

1 Robert Rynasiewicz, "Newton's Scholium on Time, Space, Place and Motion," in *The Stanford Encyclopedia of Philosophy*, 2011, https://plato.stanford.edu/entries/newton-stm/scholium.html.
2 From N. David Mermin's very readable book *It's About Time: Understanding Einstein's Relativity* (Princeton, NJ: Princeton University Press, 2009).
3 The GPS is not part of the Internet proper; it has its own radio broadcasting system for the very reason that it *does* need excellent time keeping.
4 As quoted in Viviane Richter, "A New Tool to Study Neutron Stars," *Cosmos Magazine*, June 24, 2016, https://cosmosmagazine. com/physics/a-leap-forward-in-gravitational-wave-detection.

13

QUANTUM FIELDS AND THEIR EXCITATIONS

What are classical particles and fields?

In a classical physics description of Nature, all elementary entities are considered to be either particles or fields. In this theory, a *particle* is an object that has mass—meaning it has inertia and is subject to the attracting force of gravity that exists between any two particles. A particle can also carry electric charge—meaning, it is subject to the attracting or repelling forces that exist between any two charged particles.

During the 1700s and 1800s, it bothered scientists that forces were felt between particles that were not in bodily contact, but were some distance apart. They called this uncomfortable fact 'action at a distance.' To fill the void between distant particles, physicists devised the concept of fields. In classical physics theory, a *field* is an invisible entity that permeates all of space and 'carries' a force from one object to another. For example, in this view, the Moon and Earth attract one another through the gravitational field their mass creates around them. Although the two objects are not in bodily contact, both are in local contact with the gravitational field.

On first thought, it's not clear that the idea of fields is needed. Are fields only a figment of a theorist's imagination? Physicists say no. Consider a hypothetical event in which Earth suddenly moves a little closer to the Moon. Because Earth and the Moon are now closer to each other, the mutual force of gravity is stronger. However, an important fact of Einstein's relativity theory is that no physical influence can travel faster than the speed of light, as we discussed in Chapter 8. For this reason, the effect of Earth being closer won't be felt by the Moon any sooner than it would take a pulse of light to travel between them, which is around 1.3 seconds. We can visualize that the sudden movement of Earth creates a ripple in the gravitational field near Earth, which then travels at the speed of light outward in all directions. You can visualize the ripple like a ripple on the surface of a lake after a stone is tossed into it. But there is virtually no matter between Earth and the Moon—only the gravitational field. When the gravitational ripple hits the Moon, the strength of gravity exerts a sudden and attractive force on it.

A general principle of physics is that, overall, momentum is always constant during the interaction of any group of objects. Momentum equals the velocity of an object multiplied by its mass. That is, a heavier and/or faster object carries more momentum. When a billiard ball hits another ball initially at rest, some momentum is transferred from the faster ball to the at-rest ball. Although the momentum of each ball changes, the total sum of the momenta of the balls does not change; it is constant.

This raises a puzzle. When the gravity ripple hits the Moon, the Moon is suddenly pulled toward the Earth, changing the course of the Moon's motion. That is, the Moon's momentum has changed. The puzzle that faced early scientists was: during the 1.3 seconds when the Moon does not feel any change of gravity, where is the momentum that the Moon will acquire when the gravity ripple hits it? The answer must be: the momentum resides in the gravitational field itself. This realization inspired scientists to realize that the gravitational field is more than a figment of imagination. It's an actual entity in Nature.

The same argument applies to the electric field and the magnetic field. They have momentum and energy. When an electrically charged object is moved suddenly by something pushing on it, it creates electromagnetic radiation—for example, radio waves or light. This radiation is received subsequently by an object such as a radio receiver antenna, or, in the case of light, your eyes. But, before this form of energy hits the antenna or your eyes, where does the energy reside? In the electromagnetic field!

If something carries momentum and energy, then it is a physical entity. So, in the classical physics description of Nature, there are two kinds of elementary things: particles and fields. In this picture, particles and fields are sufficient for describing Nature.

What quantum physics principle unifies the concepts of particles and fields?

All physical entities in Nature are quantum; that is, they can be described using quantum theory. This should include physical fields. The first physical field to be described successfully by quantum theory was the electromagnetic field, or EM field for short. As discussed in Chapter 1, during the 1920s, Max Born, Werner Heisenberg, and Pasqual Jordan, and separately Paul Dirac, devised a quantum theory description of the EM field, and ushered in a new era of physics.

Recall that, in a classical physics description, the EM field carries disturbances in the form of waves, or ripples. The field features oscillations as it moves at the speed of light. In fact, these waves *are* light. As an analogy, consider a classical water wave moving across a lake. It helps to visualize such a wave by imagining there are many small corks spread uniformly and floating on the lake's surface. As a wave passes by, the corks oscillate up and down in a wavelike pattern. The wave is the coordinated motion of many entities at different locations. A wave in the EM field is analogous; at every point in space, the field has a particular value or strength. An EM wave does not move up and down. Rather, the field's value oscillates from strong to weak at all locations in a coordinated manner.

To make a slightly more accurate analogy, imagine a million corks arranged in a chessboardlike grid in an area the size of a tennis court. Each cork has four closest neighboring corks, and each cork is connected to each of its neighbors by a short, taut spring. Now suspend the entire array of interconnected corks by its four corners, so no cork touches the ground, as in FIGURE 13.1. If all the corks are initially at rest, and you reach up and 'pluck' the center cork, a wave of motion will radiate out from that location. Every cork will begin vibrating or oscillating around its original location, all with the same frequency. This wave of cork motion can be described using classical physics.

Figure 13.1 A grid of connected objects set vibrating by being plucked at the center. Motion sensors are placed around the edges.

What if you replace each cork by a tiny object that is described using quantum theory? How is the wave of motion described using quantum theory? Recall that in quantum physics we can't talk about the motion of each object as we would when using classical theory; the concept of a trajectory of motion is not, generally, a meaningful one. To be clear, let's stipulate that the grid of objects is enclosed in a darkened room so you can't observe the objects' motion, and each object does not leave any permanent trace that would enable you to know it was located at a particular place at a particular time. That is, in the language of quantum measurement, assume the objects are not measured. Now, the proper quantum theoretic description of the motion of the whole grid of objects is as one unitary process. As introduced in Chapter 4, a unitary process cannot be divided into individual steps, each with a definite, observed outcome.

The whole grid of spring-connected objects acts as a single quantum system or entity. This idea goes beyond considering two objects together as a composite object, as we did earlier when discussing entangled quantum states. Now we have an entangled state involving millions of objects. The state of each object cannot be represented independently of the others. State entanglement that involves field entities at many locations is the essence of the quantum theory of a field.

What happens if we measure a quantum field?

Let's say we intend to pluck the object located at the center of the grid. The grid is in the dark and not measured. To make a measurement, we place a small motion sensor next to one of the objects at the edge of the grid. After we pluck the center object, the edge object may jiggle, setting off its motion sensor, which indicates a detection event by flashing a light bulb. This constitutes a measurement of an edge object only; the objects in the center part of the grid were not measured.

We could make this experiment more interesting by placing motion sensors at all of the edge objects. If the grid were made of objects described properly by *classical* physics, then every edge object would jiggle sometime after the center object was plucked. Let's say we repeat the plucking-and-observing experiment many times, each time reducing the strength of our plucking. Again, every edge object would jiggle, but with decreasing vigorousness. If the motion sensors were sensitive enough, each would indicate the arrival of the wave at the edge location it monitors.

If the grid is made of *quantum* objects, then a curious effect is predicted by the theory. Again, repeat the experiment many times, each time decreasing the strength of plucking. Again, we would expect that every edge object would jiggle, if ever so slightly, sometime after the center object was plucked. Born, Heisenberg, and Jordan found that this is not the case! Quantum theory predicts that when the plucking strength is very weak, only a small fraction of the edge objects are observed to jiggle and trigger their sensors. The others don't jiggle at all. As the strength of the center plucking is decreased gradually, the number of edge objects triggering their sensors decreases, but the vigorousness with which those particular objects jiggle does *not* decrease! Furthermore, if the identical experiment is tried several times, it is predicted that a different set of edge objects will jiggle on each try. It cannot be predicted which edge objects will actually jiggle on a particular try. Only the probability for each object to jiggle can be computed from the theory. If the plucking strength is reduced even further, a situation is reached in which only one edge sensor is triggered. That is, only one edge object begins vibrating, and on each repeat of the experiment, a different edge object could be observed to vibrate.

Such curious predictions can be explained tentatively by a simple model in which the weakly plucked center object 'shakes off' a small number of distinct entities called *energy quanta*, each of which contains the same amount of momentum and energy. When one of these quanta leaves the center object and hits an edge object, that object begins to jiggle and triggers its motion sensor, or detector. If it is hypothesized that the quanta are sent flying off in random directions, then it can be understood why, during different tries, different edge objects jiggle. To complete this simple model, one should also assume that the more vigorously the center object is plucked, the greater the number of quanta that will be shaken off, and also the greater the number of edge objects that will jiggle—all with the same vigorousness. In the situation in which only one *quantum* (the singular form of quanta) is shaken off, only one edge object will begin to vibrate and thus trigger its sensor.

How does the quantum theory of a grid apply to light?

Born, Heisenberg, and Jordan deduced from quantum theory that if the grid is oscillating at a single, well-defined frequency (number of cycles per second), each of the quanta created in the grid should carry the same amount of energy. The theory indicated that this amount of energy should be related to the frequency of the grid oscillation by Planck's original formula, which we met in Chapter 6:

$$Frequency\ of\ grid\ oscillation = \frac{Energy}{h}.$$

Again, h stands for Planck's constant, which is recognized as a fundamental constant of Nature. This was quite a nice result, because if we replace the grid in our thinking by the electromagnetic field, it predicts the same result as predicted by Planck and Einstein's older theory of light based on photons. Apparently, the energy quanta of the grid play the same role as photons, which, for a given frequency or color of light, have an amount of energy in accordance with Planck's formula.

Therefore Born, Heisenberg, and Jordan felt confident in applying their oscillating-grid theory to the electromagnetic field. They recognized that, instead of having a physical object located at each position in the grid, the electromagnetic (EM) field itself has a value or strength at each position in the grid. This value corresponds to how strong a force the field would exert on a charged particle if placed at that location. So they mentally removed the material objects from their grid model, leaving only the EM field. The field at each location oscillates at a certain frequency, depending on the color of the light—blue, faster; red, slower. This conceptual disappearance of the imagined material objects and the persistence of the field in its place might remind you of Lewis Carroll's Cheshire Cat, whose disembodied smile was the only aspect remaining after it disappeared.

If a charged particle such as an electron is jiggled or shaken, it gives off electromagnetic radiation. If the radiation is visible to the eye, we call it light. Let's say an electron is located at the center of a dark room when it is shaken, and that many photodetectors are located on the room's walls. If the electron is shaken very gently, then it is observed that only one detector will register an event—meaning, it received a quantum of light. It cannot be predicted which detector will actually register the event; it is random. We can calculate only probabilities for any detector to register a detection event. Instead, if the electron is shaken slightly more vigorously, then more than one detector may register an event, again following the probabilities predicted by quantum theory.

It seems natural to imagine particles called photons being shaken off from the electron and traveling to the detector, but that is a far-too-simplistic viewpoint. The more correct view is that the entire quantum EM field between the electron and the detectors becomes activated, or excited, and begins oscillating. Then, at some moment, one or more of the detectors registers a detection event and the field becomes deactivated.

What is a quantum field?

From the previous discussions there emerges a consistent understanding of what a quantum field is. We can assert the following:

A quantum field consists of an infinite number of disembodied quantum entities—each at a point in space—that may oscillate in concert in a wavelike motion, moving energy and momentum within the field.

The quantum entities are abstract, in that they represent only the value of the field at each point in the absence of any actual matter at that location. Each entity making up the field remains at its designated location, whereas energy and momentum are passed from location to location, eventually arriving at various locations where the field's energy may be absorbed, creating detection events. The value of the field at each location represents the quantum possibility that a detection event may occur at that location.

What is a photon?

From the field theory point of view, we may now say:

> A *photon* is an individual activation or excitation of the quantum electromagnetic field.

A photon is not a particle. To emphasize this point, note that an excitation of a field, like that of a grid of objects, does not exist at a single point in space. It can be spread out over a large area or it can be concentrated into a small area. In this sense, the size of a photon depends on how you make it. If an electron is shaken suddenly and briefly in just the right way, a packet of EM oscillations will move away from the electron in a tight 'clump.' If an electron is shaken for a while, an excitation of the surrounding field will occur that extends over a broad region. This means that if an electron is shaken for a long time, and it creates a photon, the photon will be spread over a broad region. A photon is definitely not a pointlike object. In fact, it is not an object at all.

Again, let's emphasize that a photon, being an excitation in a field, is quite different than an excitation of a field as described by classical physics. As explained previously, a classical excitation of a field imparts its energy and momentum to all the surrounding detectors in roughly equal amounts. In contrast, the photon imparts its energy to one detector only, and this one detector is chosen at random by probabilities that are intrinsic to quantum physics. The photon is indivisible or, as physicist Art Hobson put it, "You can't have a fraction of a quantum."[1] It is this indivisibility of a photon that is unique to the quantum theory of fields.

Are particles and fields aspects of the same thing?

In a classical physics description of Nature, the elementary entities are particles and fields. However, Born, Heisenberg, and Jordan, and others, found that in quantum theory, there are particlelike entities, such as photons, that appear to arise naturally when considering the quantum theory of an infinite number of entities or field values capable of passing energy and momentum between them. That is, the particlelike entities are merely aspects of the quantum field; they are not separate 'things' in Nature. This realization can be seen as a great theoretical unification of different aspects of Nature that were thought previously to be quite distinct.

Does the unification of fields and particles also apply to electrons?

Yes! To appreciate the discussion that follows, you must try to disavow your conception of an electron as a tiny, stonelike object that has a certain mass and electric charge. Although I have used the word 'object' for an electron many times in this book, I have tried not to imply that it really is a tiny, stonelike object. Now, let's take a leap of thinking and conceive of an electron in an entirely new way.

Just as there is an electromagnetic field, whose quantum excitations are photons, there is also an *electron matter field* whose quantum activations or excitations are electrons. The electron matter field is a kind of three-dimensional grid permeating all of space. Its excitation into an oscillating behavior corresponds to the appearance of an electron. From this viewpoint, an electron is not a particle; it is a quantum excitation of the electron matter field.

How can you 'pluck' the electron matter field to create an electron that travels away from the plucked location? The theoretical discovery of such a process—proposed by Wendell Furry, J. Robert Oppenheimer, Wolfgang Pauli, and Victor Weisskopf in the 1930s—was a landmark event in the history of quantum physics, and it led the way to our modern theories of all elementary particles. It had been known previously that the electromagnetic field interacts with the electron matter field in such a way that energy can be exchanged between the two. This is not surprising; even in classical physics theory an oscillating electron imparts energy to the EM field.

Physicists found that quantum theory predicted the EM field can interact with the electron matter field in a much more drastic way, such that a quantum of EM energy—a photon— ceases to exist, and an electron is born of the energy given up by the EM field when the photon blinks out of existence. To put this into more proper field language, the EM field loses one quantum of excitation and the electron matter field gains one quantum of excitation. As a bonus, another matter field— hitherto unknown to science—also gains one quantum of excitation. This field corresponds to antimatter—specifically, the *positron*—which has the same amount of mass as an electron, but an opposite, positive charge.

Experimental physicists indeed observed the process that quantum theory predicted. The EM field lost a photon, and both the electron matter field and the positron antimatter field gained one excitation each. The total energy during this process was constant; it just passed from one form to another. Scientists were impressed that the mathematics predicted correctly that matter, in the form of electrons, could be created and destroyed when fields exchanged energy. They were doubly impressed that the mathematics predicted correctly the existence of an entirely new kind of matter: antimatter.

Theorists were gratified that quantum theory's prediction was consistent with Einstein's relativity theory, which predicted that matter and energy are interconvertible. Einstein's famous equation $E = mc^2$ represents the fact that energy E and mass m, when interchanged, are directly proportional, whereas the speed of light c enters the relation as a constant factor. Recall it was the speed limit imposed by the speed of light that provided a convincing argument that fields are real physical entities; the fields accounted for energy and momentum during the time between an event like Earth suddenly moving and the gravitational effects being felt by the Moon. What physicists did during the 1930s was to bring together Einstein's theory of relativity and quantum theory. They found that this marriage predicted specific processes by which energy and matter could be interchanged.

Why don't we see ordinary objects appearing and disappearing?

If the theory of quantum fields predicts that matter can be created and destroyed, why are ordinary objects such as pencils and potato chips not appearing and disappearing all the time? We live in a low-energy world. The objects around us have energies that are set by the scale of room temperature or a little warmer. If most of the objects around us were much more energetic than this, we humans wouldn't be here on Earth.

The creation and destruction of elementary particles occurs only when the energies involved are very high—much greater than the ordinary energies of objects around us. Such conditions exist inside the Sun and inside high-energy particle accelerators. Both are places we would rather not be.

What is the universe made of?

Just as there is a matter field corresponding to electrons, there is a field corresponding to protons and neutrons—a proton matter field and a neutron matter field, respectively. In fact, every elementary 'particle' has a corresponding matter field, with excitations that are 'particles' of that type.

Steven Weinberg, a Nobel-winning quantum physicist, summarized the development of quantum field theory:

> Thus, the inhabitants of the universe were conceived to be a set of fields—an electron field, a proton field, an electromagnetic field—and particles were reduced to mere epiphenomena. In its essentials, this point of view has survived to the present day, and forms the central dogma of quantum field theory: the essential reality is a set of fields subject to the rules of [Einstein's] relativity and quantum mechanics; all else is derived as a consequence of the quantum dynamics of those fields.[2]

Art Hobson summed up this idea succinctly as "Fields are all there is."[3]

What is the quantum vacuum?

In prequantum days, physicists had conceived of several hypotheses about what remains in a region of space where all matter and radiation are absent. Actually, there is nowhere in the universe that is completely void of matter or energy, but that didn't stop theorists of old from theorizing. One idea was that such a region, called a *vacuum*, would contain absolutely nothing. Another idea was that it would be permeated by an 'ether,' which was thought to be a substance unlike energy or matter that would provide a kind of signpost at every point in space, telling energy and matter where it is, and which way is up, down, left, or right. This classical concept of the ether was disproved by an experiment in 1887 by Albert Michelson and Edward Morley. They used an interferometer to measure the speed of light in two perpendicular directions. The absence of any difference in these speeds is consistent with the absence of ether, and also is consistent with Einstein's theory of relativity.

So, what is a vacuum? It's not nothing. The quantum theory of fields says all of space is permeated by quantum fields—an electron field, a proton field, an electromagnetic field, along with others corresponding to each kind of matter. Even when there are no excitations or 'particles,' the fields are still present. We call such an otherwise empty region the *quantum vacuum*.

What are the properties of the quantum vacuum and how does the vacuum make itself felt? An important aspect of the quantum vacuum can be realized by applying the principles of quantum physics we discussed in earlier chapters. First, any quantum entity does not have predetermined measurement outcomes; only quantum possibilities exist before a measurement. Second, certain quantities are complementary; they cannot be measured precisely and simultaneously. This idea leads to Heisenberg's Uncertainty Principle. That is, if a particular quantity is measured, then another quantity is indeterminate. For example, if a particle's position is measured, then its velocity, or rate of change of position, is indeterminate. The question "What is its velocity?" has no meaning in this case.

Applying the Uncertainty Principle to a quantum field— say, the electromagnetic field—leads to a remarkable fact. It is *impossible* for the quantum field to be perfectly quiescent—that is, perfectly at rest or inactive. If the value of the field at a certain location were precisely zero, then its rate of change would not be zero. So, after a brief instant of time, its value would no longer be zero! The best that can be done is to remove all particlelike excitations (quanta) from the field; then, its value and rate of change of value can both be at minimum values, but neither is zero. If we were to measure either one, we would obtain a nonzero value that is random, according to probabilities calculated using quantum theory.[4] The average of these random values is zero. These random values are referred to as vacuum fluctuations.

We can conclude that the quantum vacuum is filled with fluctuating quantum fields, which are crudely illustrated in FIGURE 13.2. The fluctuating fields take on definite values only when being measured or when interacting in an irreversible manner with other fields. The first experimental evidence for this prediction of quantum field theory was found by careful measurements of the (quantized) energies of electrons in the hydrogen atom. A quantum theory ignoring the existence of the vacuum fluctuations failed to predict the energies correctly, whereas a theory that accounts for the vacuum fluctuations does predict the correct energies to better than one part in a million! The impressive agreement between experiment and theory convinced physicists of the reality of the quantum vacuum, and it earned Willis Lamb a Nobel Prize in 1955.

(i)   (ii)

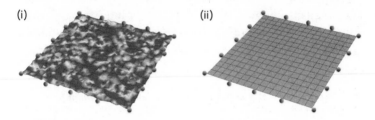

Figure 13.2 (i, ii) Quantum vacuum (i) and classical vacuum (ii).

How did the elementary particles get their mass?

Another, very important, example of physical consequences of quantum vacuum fields involves the Higgs field vacuum. Although first predicted to exist by Peter Higgs during the 1960s, the Higgs field went undetected in experiments until 2012—an event that garnered huge media coverage and a Nobel Prize for Higgs, along with François Englert. The *Higgs field* is a quantum field that is known famously for imparting mass to elementary particles, such as electrons and quarks. From a quantum field viewpoint, when the electron field becomes excited so a single electron is created, the electron encounters the all-pervasive Higgs field as it travels from place to place. The Higgs field acts against the electron's motion, kind of like water resists an object moving through it. This interaction acts to resist the acceleration the electron would experience if a separate force acts on it. This resistance to acceleration is the essence of inertia. So the picture is that the interaction of the electron with the Higgs field gives inertia to the electron, and this inertia can be thought of as arising from mass.

An interesting feature of this story is that the Higgs field remains in its vacuum state nearly all the time. Normally, there are no Higgs particles—the famous *Higgs bosons*—existing as freely moving particles. Such particles, which are the excitations of the Higgs field, are few and far between, and they don't play a direct role in imparting mass to other particles.[5] This is a powerful example of the reality and importance of vacuum states of quantum fields. Without the Higgs vacuum, and its quantum fluctuations, particles would weigh nothing! All particles would travel at the speed of light, like photons, and there would be no atoms. And no us.

What other facts speak in favor of the existence of quantum fields?

It took a long time, from the 1930s to the 1960s, for physicists to accept fully that quantum fields really exist in Nature. Even today, quantum field theory is treated mostly as an esoteric, advanced part of quantum theory, reserved only for experts. Nevertheless, many experts, such as Weinberg and Frank Wilczek, are of the opinion that quantum fields are more fundamental than particles, and that this viewpoint should be more widely appreciated. The following arguments support this supposition:

- **Quantum fields 'think globally,' but act locally.** Wilczek writes, "The concept of locality, in the crude form that one can predict the behavior of nearby objects without reference to distant ones, is basic to scientific practice."[6] Quantum field theory satisfies this desire and describes successfully all phenomena to which it has been applied, without invoking action at a distance, which would violate Einstein's theory of relativity. This includes describing correctly the counterintuitive Bell-test correlations that can be observed when measuring correlated objects that are separated by a great distance.

• **Quantum fields evince identical particles.** Wilczek writes, "Undoubtedly the single most profound fact about Nature that quantum field theory uniquely explains is the existence of different, yet indistinguishable, copies of elementary particles."[7] The fact that the world is made of a limited number of particle types, and that any two members of the same type are identical, is by no means obvious. For example, any two electrons are identical—that is, completely interchangeable. "We understand this as a consequence of the fact that both are excitations of the same underlying 'ur-stuff,' the electron field. The electron field is thus the primary reality," Wilczek says.[8]

• **Quantum fields account naturally for changing numbers of particles.** Quantum field theory not only accounts for the creation and destruction of photons when atoms emit or absorb light; it also accounts for processes such as the creation and destruction of electrons and positrons. Wilczek writes, "In this picture it is only the fields, and not the individual objects they create and destroy, that are permanent."[9]

• **Entanglement can exist with only one particle present.** In my previous descriptions of entangled quantum states, I talked about two 'objects'—for example, photons—with a combined state of the form (0)&(1) + (1)&(0). Here, I am using the language of qubits, where (0) and (1) refer to any two states that can be used to characterize each object. Recall, the symbol '+' means 'in superposition with.' What if we now consider two distinct regions of the electromagnetic field? Let's say we have two microwave ovens separated by five kilometers. We could prepare each oven so it contains either zero or one microwave photon's worth of quantum field excitation. We label the state of the EM field inside each oven by either (1), if it has one photon's worth of excitation, or (0), if it has none. Now the state (0)&(1) + (1)&(0) refers to the state of the two EM fields. This is an entangled state of the two ovens' fields. It represents a situation in which each oven's field has a possibility to be excited to contain one photon's worth of energy, but there is no possibility that both fields are excited. (It is not an entanglement of states of particles.) Such field entanglement can be verified with laboratory experiments in which the two ovens remain distant and no quantum entities are exchanged between laboratories. Quantum physicist Steven van Enk writes, "I would conclude that the state (0)&(1) + (1)&(0) [has] entanglement."[10] The fact that two separated EM fields can have entanglement, even though there is only one photon's worth of excitation between them, implies that EM fields are truly physical entities.

- **Quantum field theory gives a clearer picture of wave–particle duality.** The electron matter field is not an electron. Rather, an electron is an individual excitation of the electron matter field. The electron matter field itself behaves in a wavelike manner, and it represents possible measurements to determine where an electron is most likely to be found. Therefore, it is not surprising that, if one believes mistakenly that an electron is a particle, apparent contradictions and meaningless questions can arise. For example, the question "Which path did the electron take on its way to a detector?" has no meaning. On the other hand, a quantum field permeates all of space; therefore, it exists within both paths. So the proper statement is not that an electron sometimes behaves like a wave and sometimes like a particle. Rather, one should say, the quantum field always behaves like a quantum field with its wavelike behaviors, and the electron is a manifestation of that field. It is best to replace the mysterious concept of 'wave–particle duality' by the less mysterious concept of 'quantum field–quantum particle duality.'

Does an understanding of quantum fields remove the mystery of Bell correlations?

No, it does not. Recall that the awkwardness of the nonlocal Bell-type correlations of measurements of distant objects has nothing directly to do with quantum physics. So it doesn't matter whether you consider quantum particles or quantum fields as more fundamental. The awkwardness remains and the worldview called Local Realism is still untenable.

Does an understanding of quantum fields remove the mystery of quantum measurement?

No, it does not. The mystery of quantum measurement, stated in terms of quantum fields, is the following: Let's say an atom has its electron in an energetically excited state. Then, it emits light in the form of an excitation of the quantum EM field: a photon. Now the atom has reduced energy and the field has increased energy. The energy created in the field travels out to a set of equally distant detectors. Let's say Alice is stationed at one detector and Bob far away at another, and Alice sees her detector register an event. She now knows that all the photon's energy has been deposited in her detector, because the photon is indivisible. And she also knows that Bob's detector cannot register any energy from the atom's giving up of energy.

This sounds reasonable, but there is a catch: If the ripples that travel away from the atom in all directions carry energy locally in each ripple, then how does all the photon's worth of energy suddenly become localized in Alice's detector? Clearly, energy cannot jump instantaneously across a great distance from one region to another region.

To understand this situation requires subtle thinking, and forces us to consider further the meaning of a quantum field. Rather than being like a classical field, in which energy resides locally in each portion of the traveling ripples, a quantum field does not represent any actualized physical reality. Rather it represents only quantum possibilities. This explanation is a generalization of those I offered in earlier chapters when talking about a single photon traveling through two possible paths to a detector. We had to be careful not to attribute an actual reality to each possibility. We had to distinguish between quantum possibilities and measurement outcomes. Therefore, as we did for a single electron, we need to keep in mind that the quantum state of a field represents only possibilities and is not in one-to-one correspondence with measurement outcomes. It might be hard to visualize a quantum field as a grid of traveling quantum possibilities, but this is perhaps the closest description of a quantum field I can devise.

Why is the discussion of quantum fields postponed to near the end of this book?

As it also happened historically, the concepts behind quantum fields cannot be explained until after the general principles of quantum physics are understood. Most of this book is devoted to this task. Then, by postulating the existence of fields, and insisting they are governed by quantum principles, the powerful and beautiful quantum field theory emerges, along with all of its correct predictions about nature.

Notes

1 Art Hobson, "There Are No Particles, There Are Only Fields," *American Journal of Physics* 81 (2013), 211–223; quote, 214.

2 As cited in Heinz R. Pagels, *The Cosmic Code* (Mineola, NY: Dover, 1982), quote, 269.

3 Hobson, "There Are No Particles," 214.

4 My research group at the University of Oregon carried out such measurements in 1994 using quantum-state tomography to measure the quantum state of the vacuum field.

5 The experiment in 2012 managed to create enough Higgs particles to detect them by smashing together high-speed protons

in a particle accelerator called the Large Hadron Collider, located in Switzerland.

6   Frank Wilczek, "Quantum Field Theory," *Reviews of Modern Physics* 71 (1999): S85–S95; quote, S85.

7   Wilczek, "Quantum Field Theory," S86.

8   Wilczek, "Quantum Field Theory," S86.

9   Wilczek, "Quantum Field Theory," S87.

10 Steven van Enk, private communication, 2016.

14

FUTURE DIRECTIONS AND REMAINING QUESTIONS IN QUANTUM SCIENCE

What is needed to make further progress?

To make further progress, quantum scientists need to understand better the nonintuitive aspects of quantum phenomena and their description by quantum theory, and they need to develop new technologies for building devices that rely on such phenomena for their operation.

Quantum physicists are working to understand in deeper ways the nonlocal correlations that can occur when making measurements on quantum-entangled objects. The Bell-test experiments discussed in Chapter 8 demonstrate that such correlations cannot be explained using the classical physics concepts of local causality and preexisting properties of objects, ruling out the conceptual framework called Local Realism. Quantum theory *can* describe the observed nonlocal correlations, using the concept of entangled states of two objects. Such entanglement can occur even though two objects are separated in space by a great distance. The quantum description makes it clear that what happens to one particle in no way affects directly the state of the other particle; yet the correlations still occur in a manner that defies classical physics description. Physicists would like to know how such correlations occur across large distances. Although quantum theory describes these correlations perfectly, it doesn't say *how* they come to be. Is there a 'backdoor' channel that somehow transmits the correlations without violating the cause-and-effect nature of things that seems to be required by relativity theory? What becomes of the nonlocal correlations when the two quantum objects are near or inside a black hole, where space–time becomes strongly warped? In such regions the concepts of time and space require us to rethink what we mean by 'local.' Questions such as these might lead to breakthroughs in understanding the nature and early history of the universe itself.

Quantum technologists are working to increase their skills in device design and construction needed to advance the three main areas of research and development: quantum communications, quantum sensors, and quantum computing. According to a report released recently by two professional societies,

A global quantum revolution is currently underway. . . . This revolution is driven by recent discoveries in the new area of quantum information science, which is based on the recognition that the subtler aspects of quantum physics such as quantum superposition and entanglement are far from being merely intriguing curiosities and can be transitioned into valuable, real-world technologies. Quantum science and technology will revolutionize many aspects of our lives, including improved security and privacy in digital communications systems that connect our world; enhanced navigation in demanding environments; advanced sensors for geological resource exploration; and superior computational capabilities for complex simulations and modeling of new pharmaceutical drugs and solar-energy-harvesting materials.[1]

Technological breakthroughs are needed, for example, in the following areas: The first includes sources of light that reliably create single photons at known, controllable times. The challenge is to overcome the randomness of quantum outcomes in producing such photons. The second breakthrough required involves compact, portable atomic interferometers for inertial sensing. These devices rely on the ability to cool a small, confined cloud of atoms to temperatures near absolute zero. Third, methods for constructing a 'scalable' quantum computer—that is, a scheme in which doubling the size of the computer memory and processors requires only twice the cost and space (not an exponential increase in them). An example of the latter is the need for improved techniques for placing single atoms at known locations in a solid material such as silicon or in a magnetic trap in an evacuated chamber, and the means for manipulating and probing their quantum states. All of these and more are the subjects of intense, ongoing research.

What don't we know about quantum technology?

We don't yet know how far quantum technology can be pushed to create alternatives to classical physics–based technologies with enhanced capabilities. Will reliable, working quantum computers be built successfully and, if so, what tasks will they be used for most productively? Will quantum-based sensor technologies mature and be deployed in a wide range of applications? Or will the complexity of these devices and their extreme sensitivity to small disturbances make them uneconomical?

I say, "Don't bet against the technologists." There are no known barriers—from a physics point of view—to building these things. A path will be found sooner or later to harness the unusual, nonintuitive behaviors of quantum systems to create these or other still-unimagined devices and technologies.

The new technologies will likely not replace existing ones but will augment or supplement them, and each will be used where it is best suited. This prediction parallels the idea that both classical physics theory and quantum theory have their places and their roles in describing physical systems. We use the one that is best suited for solving a particular problem at hand.

What don't we know about quantum physics?

To make progress in science, it is most important to know what you don't know. Asking the right questions is paramount. You might still find yourself puzzling over some of the aspects of quantum phenomena and their theoretical description that were discussed throughout this book. If so, welcome to the crowd. Nobel Prize winner Murray Gell-Mann said:

> The discovery of quantum mechanics is one of the greatest achievements of the human race, but it is also one of the most difficult for the human mind to grasp. ... It violates our intuition—or rather, our intuition has been built up in a way that ignores quantum-mechanical behavior.[2]

So let's explore a little more deeply some of the questions that might have arisen in your mind while reading this book. First, let's review what we *do* understand about quantum physics.

What do we understand about the quantum aspects of Nature?

Above all, we have learned that Nature is probabilistic—that is, some events occur in an intrinsically random way. For example, if an electron is excited to a high-energy state in an atom, it will decay to lower energy, emitting a photon. It can decay at any time after excitation, and there is simply no way to predict precisely when this will happen. On the other hand, causality is still upheld; a particular event cannot occur unless the necessary prior events have occurred and the necessary prior conditions have been met. Given certain prior events and conditions, there is more than one possible future, each with its own probability of occurring, but each has to be consistent with the ideas of cause and effect.

Again I quote Gell-Mann, who said, "The description of the universe in quantum mechanics is a probabilistic one. The fundamental laws do not tell you the history of the universe. They tell you probabilities for an infinite set of alternative histories of the universe."[3] This description is different from the one provided by classical physics theory, which says that if you know the state of everything at the present time, you could, in principle, predict perfectly the state at all later times.

How does quantum theory represent the unfolding of physical history? Recall, Born's Rule tells us that the probability of an event is given by the square of the 'possibility' for that event. And recall, possibility is the component of a state arrow pointing to a certain outcome for the event. The Schrödinger psi wave represents an infinite collection of such possibilities. When a measurement outcome is recorded or, more generally, when an event has happened irreversibly, the state arrow or the psi wave becomes altered. This altering or 'updating' comprises the specific new conditions that allow for a range of possible future events.

So the behavior of the world is causal, in that not 'any old thing' can happen, but it is also probabilistic, in that a given present can lead to a range of possible futures. Although there are mysteries regarding how the theory should best be interpreted, the rules for using quantum theory to compute the probabilities of possible futures are clear, and, as this book illustrates, are very useful in devising new technologies.

How do the classical and quantum descriptions of Nature differ?

From a classical physics viewpoint, our inability to predict the future arises only from our lack of complete knowledge of the current state of everything and our inability to calculate Newton's equations perfectly. In that view, the need for probabilities is reducible to a more fundamental set of causes or reasons. Call this viewpoint *reducible probability*.

In contrast, probability in quantum phenomena is believed to be *irreducible*—that is, not arising from any underlying hidden mechanism or lack of knowledge, but innate to the world at a fundamental level. Random or probabilistic behavior is an aspect of Nature, not merely of our attempts to understand Nature. In quantum theory, the quantum state is postulated to be the most complete description of a physical situation possible. In this view, there can be no missing information 'out there' that could be acquired. This conclusion has been arrived at not by trying to predict the most complicated system—the universe—but by trying to predict the simplest systems, such as single electrons or photons. Experiments and logical thinking led physicists necessarily to describe phenomena involving these entities using quantum states, along with their attendant peculiar features: unitary evolution, superposition, and entanglement. I hope the many examples and discussions in this book have offered some understanding of these features.

What challenges remain in understanding quantum theory?

Currently, we have the existing quantum theory, which covers nearly everything we know about: classical and quantum. Yet we still don't have a perfectly clear picture of what quantum theory is trying to tell us about Nature, or at least not one that is agreed on by a large majority of physicists. There is a community of physicists trying to come to grips with the underlying reasons *why* observed phenomena are best described by quantum theory and the form it takes in terms of state arrows, superposition, unitary processes, and so on. While agreeing that the theory we now have 'works,' and allows the design and construction of new quantum technologies, these physicists hope that exploring the underlying basis of quantum theory at its most fundamental, even philosophical, level will lead to further breakthroughs. If you are interested in such questions, read on!

What is the measurement problem?

To many physicists, an apparent 'problem' with the formalism of quantum theory (state arrows, superposition, and unitary processes) is that there seems to be no clear way to place a dividing line between measurement processes and the so-called unitary processes, which occur when no measurement is taking place. An example illustrates this: An electron is emitted from a hot metal wire inside a chamber containing no air. The electron has equal possibilities to travel in any direction upon leaving the wire. The inside walls of the chamber are covered with a large number of detectors, only one of which can register the electron arriving. Because the chamber contains no air or anything else that could detect the electron on its way from source to detector, quantum theory describes the behavior of the electron before being detected as a unitary process, and represents all the possibilities using the de Broglie–Schrödinger psi wave. The psi wave spreads out from the source until it encounters the many detectors in the chamber. The wave has nonzero values at the locations of every detector. The psi wave represents a superposition of possibilities for any one detector to register the electron and not the other detectors. Then one detector—you can't predict which—registers the electron, after which no other detector can do so. This means the value of the psi wave is now zero at the locations of all other detectors. All the energy and mass associated with the electron has been concentrated suddenly at the one detector.

A puzzling feature of quantum theory is that there seems to be no physical mechanism that causes the psi wave to 'collapse' from nonzero values to a zero value at all the other detectors. In fact, if there were such a mechanism, its physical influence would need to act in a way that would violate Einstein's relativity theory; the effect of a detection event at one detector would need to 'reach out' instantaneously to affect the values of the psi wave at distant locations. There is no mechanism in the theory for such an effect, and therefore virtually no physicists believe faster-than-light 'collapse' of the psi wave to be a useful picture. Instead, the psi wave is viewed, not as a physical thing that may or may not collapse, but as a mathematical element of probability theory, which depends on the information at hand being used to describe it. When new information is gained, it is as if the psi wave collapses.

One approach to thinking about this so-called measurement problem is to adopt the stance formulated by the founding fathers of quantum theory in Copenhagen during the 1920s. Werner Heisenberg explained his view of this idea:

The observation itself changes the [psi wave] discontinuously; it selects of all possible events the actual one that has taken place. Since through the observation our knowledge of the system has changed discontinuously, its mathematical representation has also undergone the discontinuous change and we speak of a "quantum jump." ... Therefore the transition from the "possible" to the "actual" takes place during observation. If we want to describe what happens in an atomic [unitary] event, we have to realize that the word "happens" can apply only to the observations, not to the state of affairs between two observations. It applies to the physical, not the psychical act of observation, and we may say that the transition from the "possible" to the "actual" takes place as soon as the interaction of the object with the measuring device, and therefore the rest of the world, has come into play; it is not connected with the act of registration of the result in the mind of the observer. The discontinuous change in the [psi wave], however, takes place with the act of registration, because it is the discontinuous change of our knowledge in the instant of registration that has its image in the discontinuous change of the [psi wave].[4]

Heisenberg is saying that a physical event happens in the physical world, independent of any human observation, and that measuring devices or other physical systems can serve to make 'observations'—meaning, permanent physical traces are imprinted on the surroundings. In between such 'observations,' a process is unitary and cannot be said to 'happen.' The change in the psi wave does not take place until the person who is using that psi wave as a predictive tool makes note of the physical event's outcome.

A classical analogy to this story is how you think about the money that appears and disappears in your bank account. The changes in the amount of money can happen independently of you, but you don't usually update your knowledge of the account balance until the end of the month, when you look at the numbers. In this view of quantum theory, the psi wave is a kind of bookkeeping method to keep track of what is happening in the world. But of course, it is much more than that because the quantum rules for updating it also encapsulate the deepest physical description of Nature that we know.

How can an entangled state be updated?

A more sophisticated example illustrating Heisenberg's viewpoint is illustrated in FIGURE 14.1. Two photons are prepared in the polarization-entangled Bell State, $(\uparrow)_L \&(\rightarrow)_R + (\leftarrow)_L \&(\uparrow)_R$, in which the order of the parentheses labels the 'left' (L) and 'right' (R) photons. Let's say you and Alice both know this state perfectly. Now Alice measures the 'left' photon using a polarization measurement scheme with the possible outcomes $(\uparrow)$ or $(\leftarrow)$. Either could happen with fifty-percent probability. Let's say Alice observes $(\leftarrow)$, and writes the outcome on paper, but does not inform you of the outcome. You know such a measurement has been performed, and because the measurement must have yielded some specific outcome, which you assume is permanent, you believe the quantum state of the 'right' photon is now either $(\rightarrow)$ or $(\uparrow)$, but you don't know which. The entangled superposition state has been 'destroyed' by the measurement, which has yielded a permanent record of the outcome. For the 'right' photon there is no state of definite polarization. Now you and Alice are in different situations regarding information about the quantum state of the 'right' photon: she knows it; you don't.

Finally, Alice tells you, "By the way, the left photon I measured had outcome (↑)." Now you know with certainty that the 'right' photon has quantum state (→). You now update your psi-wave description of the 'right' photon, and you could use this quantum state to predict accurate probabilities for the outcome of a subsequent polarization measurement on this photon using any scheme—for example, a measurement with possible outcomes (↗) or (↘).

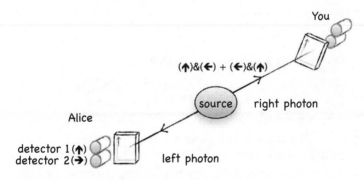

Figure 14.1 Entangled photons travel to Alice and you. If Alice measures first, what can you say about the state of your photon?

This example illustrates that, in quantum theory, both reducible (knowledge-limited) probability and irreducible (physically innate) probability can play roles. Before the measurement, you knew the combined state of the pair of photons perfectly; the only probability concept at work was the irreducible quantum probability. After Alice announced she had measured one photon but neglected to tell you the result, both types of probabilities played a role from your perspective— *reducible*, because you didn't know Alice's outcome so you didn't know the state of your unmeasured photon; and *irreducible*, because you knew your unmeasured photon could now be described by one or another quantum state, each of which implies irreducible probabilities. After Alice informed you of her outcome, you again knew perfectly a quantum state describing your photon, so only irreducible probabilities would then be involved in predicting any subsequent measurements.

Does Heisenberg's view solve the measurement problem?

Although Heisenberg's view—that the change in the psi wave happens in our thinking as a result of a measurement—is quite acceptable and is shared by many physicists, it still leaves open awkward questions. For example: How can we know what constitutes a measurement? That is, what is special about a so-called measurement device, such as a photodetector, such that it performs a measurement, whereas a device such as a calcite crystal that splits a beam of polarized photons into two beams does not, by itself, perform a measurement? As physicist Caslav Brukner writes, "At least manufacturers of photon detectors should know the answers to these questions, shouldn't they?"[5]

Niels Bohr made some progress during the 1920s and '30s toward explaining the physics behind detector operation by arguing that any true measurement device must create an 'irreversible act of amplification.' The need for amplification seems rather obvious because any useful recording of an outcome must involve a large enough number of atoms to make a permanent mark or trace that a human can see. This requires amplification from the microscopic level to the macroscopic level.

But there is a deeper reason why Bohr's statement is insightful, and that is in the concept of an *irreversible process*. As I pointed out in earlier chapters, a unitary quantum process is reversible. That is, you start with some known state of a collection of particles and then, after the unitary process takes place, you could—if you knew the state of all the particles—reverse all their velocities, pass them backward through the same process, and end up with the same state you started with. Reversibility is a hallmark of unitary processes. On the other hand, Bohr said the detection process should be irreversible to be considered a true measurement. This implies the detection process cannot be unitary. But, in accordance with quantum theory, we believe that all physical processes, if treated in sufficient detail, should be seen as unitary. This seems to imply that the measurement process itself is outside the domain of quantum theory! Are we back where we started?

A big question is: How do we go theoretically from the principles of quantum theory, which describes only unitary processes, to describing correctly a measurement process, which at least *appears* to be not unitary? A satisfying answer suggests itself: Although all processes are reversible in principle, they cannot actually be reversed in practice under the conditions in which we perform actual measurements. A unitary process can be reversed only if you know perfectly the quantum state of the material in which the measurement outcome is recorded, so the state of this material can be reversed as well. For example, the ink on the paper where Alice wrote the outcome result. There are far too many ink molecules, and their behavior is far too complex for all their states to be known precisely! This requirement to know the state of the macroscopic material stands separate from the extremely difficult task of actually reversing all the velocities of all the atoms and electrons in the material.

Here, the idea of quantum state *entanglement* shows why a true measurement cannot be reversed. Consider again the entangled photon pair shown in FIGURE 14.1. When the 'left' photon makes itself felt at a particular detector, its energy is absorbed and it triggers an avalanche of electrons stored in the detector in an unstable state, much as a skier can trigger an avalanche of unstable snow. The one photon leads to an avalanche of, say, one thousand electrons. This is the amplification Bohr talked about. If we knew perfectly the state of all the electrons before and after the avalanche process, could we write the exact state of all electrons after the avalanche? The process might be described by a state evolution something like

$$\left(1000 \; unstable \; e\right) \Rightarrow \left(1000 \; avalanched \; e\right),$$

where e stands for electrons and the arrow stands for 'goes to.'

Recall there are two detectors monitoring the 'left' photon. Let's say an avalanche at detector 1 corresponds to the outcome (↑), and an avalanche at detector 2 corresponds to the outcome (←). Then we might write the state evolution as

$$\{(\text{⬆}) \& (\text{➡}) + (\text{⬅}) \& (\text{⬆})\} \& \left(1000 \; unstable \; e\right)_{det\,1} \& \left(1000 \; unstable \; e\right)_{det\,2} \Rightarrow$$

$$(\text{⬆}) \& (\text{➡}) \; \& \left(1000 \; avalanched \; e\right)_{det\,1} \& \; \left(1000 \; unstable \; e\right)_{det\,2}$$
$$+$$
$$(\text{➡}) \& (\text{⬆}) \; \& \left(1000 \; unstable \; e\right)_{det\,1} \& \; \left(1000 \; avalanched \; e\right)_{det\,2}.$$

The original state has entanglement of the photon states (indicated in the first line by the plus sign between the polarization states), but no entanglement of the detector electron states. After the detection process takes place, the two possibilities for the 'left' photon's polarization are 'written' into the states of the electrons in each detector. According to this state evolution, the one thousand electrons in detector 1 are quantum entangled with the one thousand electrons in detector

2 (indicated by the plus sign between the second and third lines). In the second line, only the electrons in detector 1 have avalanched, whereas in the third line, only the electrons in detector 2 have done so. Both possibilities are present in the state, yet in the end only one of them can correspond to an event that has 'happened.'

The state as written still omits a crucial aspect of the measurement process. It relies on the false assumption that the detectors are perfectly isolated from the rest of the world. This cannot be so if we want them to serve as true measurement devices. The information about the avalanches has to be 'read out' in a permanent form so the information implied by the measurement can be accessed by humans, or at least computers acting as observers. The readout could be in the form of a paper printout or numbers in a computer memory. Let's call this permanent readout the 'memory.' The memory becomes quantum entangled with the photons' states and with the thousands of electrons in the two detectors. This entanglement could be represented by appending a 'memory state' to the previous equation. It would read (detector 1 registered) or (detector 2 registered), depending on which detector registered the avalanche event.

Now recall that an entangled state represents as complete knowledge as possible about the combined system, treated as a whole entity. If one part of the system is made unavailable for any reason, the remaining part cannot be described by any perfectly known state. For the situation we are discussing, the photon being measured along with the two thousand detector electrons make up one part, and the memory is the other part. Consider a situation in which the information recorded or written in the memory is 'stable'—that is, it has been present long enough to be observed by many people or copied to many electronic devices. Then, there simply is no way to know perfectly the quantum state of all those memories holding all those copies; they are far too large and complex. Just knowing the result of the measurement (say, detector 1 registered an event) is not sufficient to specify the detailed state of millions of particles, which is highly entangled. Therefore, it is not possible to reverse perfectly the whole detection process and end up back at the initial state that existed before detection. This reasoning helps in understanding how a measurement can truly 'happen' irreversibly, even though, at the heart of it all, is a collection of unitary processes.

How does decoherence help?

During the past thirty years, a detailed theory has been developed to back up the argument just presented. It is based on the idea that the measurement process effectively destroys the initial coherence of the state superposition present in an entangled state. This observation has been given the catchy name *decoherence*. A principal developer of this theory is physicist Wojciech Zurek. As he says, "Decoherence destroys superpositions." In 2002, Zurek wrote about this theory:

> The natural sciences were built on a tacit assumption: Information about the universe can be acquired without changing its state. The ideal of "hard science" was to be objective and provide a description of reality. Information was regarded as unphysical, ethereal, a mere record of the tangible, material universe, an inconsequential reflection, existing beyond and essentially decoupled from the domain governed by the laws of physics. This view is no longer tenable. Quantum theory has put an end to this Laplacean dream about a mechanical universe. Observers of quantum phenomena can no longer be just passive spectators. Quantum laws make it impossible to gain information without changing the state of the measured object. The dividing line between what is and what is known to be has been blurred forever. While abolishing this boundary, quantum theory has simultaneously deprived the "conscious observer" of a monopoly on acquiring and storing information: Any

correlation is a registration, any quantum state is a record of some other quantum state. When correlations are robust enough, or the record is sufficiently indelible, familiar classical "objective reality" emerges from the quantum substrate.[6]

Where Zurek mentions a Laplacean dream, he is referring to Pierre de Simon de Laplace's eighteenth-century classical physics theory, which assumed that, given enough information, the detailed future of the universe could, in principle, be predicted perfectly. As we have learned through our study of quantum theory, this is no longer thought to be possible.

Is decoherence sufficient?

Some scientists claim that the argument using the ideas of decoherence is not a solution to the measurement problem. They argue that, in principle, we could know the highly entangled state after the measurement process, including the state of the millions of particles making up the memories, and, in principle, we could reverse all the particles and go back to the initial state that existed before the measurement. That is, they argue that the whole process really is unitary and reversible, according to quantum theory. So, they argue, quantum theory does not describe actual, irreversible measurements properly.

A possible answer to this objection is the following: We have to decide what we mean by 'measurement.' If we just mean an amplifying process that proceeds according to quantum mechanics, then, according to quantum theory, any such process can be reversed perfectly. Although such reversal may be impractical, the theory allows it in principle. On the other hand, what if we *define* measurement as a process that leaves a permanent mark or trace in a memory? As Caslav Brukner writes, "Measurements have to result in irreversible facts; otherwise, the notion of measurement itself would become meaningless, as no measurement would ever be conclusive."[7] And physicist Otto Frisch, wrote in 1965:

> Perhaps the most important conclusion [is that] any reversible process can be reversed, given enough ingenuity. The conclusion is that a measurement is not done until some irreversible process has taken place. ... To measure is to create information, which is a state—in a machine or organism—which extends from a certain time into the future.[8]

Let's note further that a permanent mark can be copied, and that many copies can be made and distributed to our friends' memories. That is, we can create information, and record it and copy it permanently. The fact that the information gained from a measurement can be copied without error is consistent with considering it to be 'classical.' (Recall that the 'no-cloning principle,' which prohibits copying quantum states, discussed in Chapters 2, 7, and 9, does not rule out copying classical states.)

The very act of creating or gaining information and 'protecting' it permanently, by making many copies, will prevent us from being able to reverse the measurement process. This explanation of why the measurement process in not unitary doesn't violate or break any principles of quantum theory. It is a *logical* argument: IF you insist on having a permanent record of your measurement, THEN the measurement process is necessarily not reversible and, therefore, by itself, is not unitary.

Is quantum probability personal?

Heisenberg wrote, "Certainly quantum theory does not ... introduce the mind of the physicist as a part of the atomic event."[9] Yet he also said in a previous quote that the discontinuous change in the psi wave takes place with the act of registration in the mind of the observer. This raises the questions: What is *subjective* or *objective*? A dictionary definition of *subjective* is "relating to the nature of an object as it is known in the mind as distinct from a thing in itself." This is the opposite of *objective*, which is defined as "belonging to the object of thought rather than to the thinking subject."

To clarify the use of these concepts in physics theory, it is important also to distinguish between *randomness* and *probability*. **Randomness** means happening with no underlying cause or reason, and with no discernible pattern. Randomness means different things in classical theory and quantum theory. There is no true randomness in the classical physics view of Nature, at least in principle. On the other hand, in the quantum physics view of Nature, there is true, innate randomness of physical events. **Probability** is a number representing your confidence that a given event will happen (or has happened in the past). In both classical theory and quantum theory, probability is subjective, according to the viewpoint I am taking here. It is a conceptual method that thinking beings use to decide their degree of confidence about predicting possible outcomes or events.

In quantum physics, randomness exists in Nature and probability exists in the mind of the observer. This view is consistent with the view of physicist Eugene Wigner, who wrote in 1967, "the [quantum state] is only an expression of that part of our information concerning the past of the system which is relevant for predicting (as far as possible) the future behavior thereof."[10]

Closely related ideas were formalized and put forward in detail in the 2000s by Carlton Caves, Christopher Fuchs, and Ruediger Schack in a viewpoint they called the **Bayesian approach to quantum theory**. They wrote, "In the Bayesian approach to quantum mechanics, probabilities—and thus quantum states—represent an agent's [person's] degrees of belief, rather than corresponding to objective properties of physical systems."[11]

Thomas Bayes, in the early 1700s, proposed a mathematical relation between observed outcomes and the probability that those outcomes arose from a particular cause. Simon Pierre Laplace, slightly later, arrived at the same formula independently, and went further, arguing that probability should be viewed as subjective—that is, as a method of thought. He emphasized a personalized view of probability in that different people may observe the same facts but decide on different probabilities for the possible causes of those facts. Laplace's concept is that probability is a number representing the degree of confidence one has that a particular event will occur, *given* a person's prior degree of confidence, combined with any new information that motivates an update of the probability. Nowadays this is called the 'Bayesian' approach to probability theory.

In the modern Bayesian approach to quantum theory, one recognizes a quantum state as representing the information needed to determine one's degree of confidence about a certain fact or prediction. How is this concept related to 'prior information?' Let's say you prepared a state by a trusted experimental method. According to the Bayesian viewpoint, your prior knowledge about the apparatus and methods used plays a role in your assignment of the state you prepared. A concrete example of updating a quantum state upon gaining new information can help make this point clear, as described next.

Let's say the mayor tells you (and you believe her) that your and Bob's pair of photons were prepared in the entangled Bell State, $(\uparrow)_Y \&(\rightarrow)_B + (\leftarrow)_Y \&(\uparrow)_B$, where Y labels your photon and B labels Bob's. From this information you can predict correctly that whatever scheme you use to measure your photon's polarization, you will have a fifty-percent probability to observe either of the two possible outcomes: $(\uparrow)$ or $(\leftarrow)$. Then Bob tells you, "I measured my photon and found it to be $(\uparrow)$." Now you should update your predictions using the rules of quantum theory; that is, you now believe your photon to be described by the state $(\leftarrow)$, from which fact you can predict the probabilities in any measurement scheme you might choose.

Now, what if, instead, the governor tells you, "The mayor is correct that whatever scheme you use to measure your photon's polarization you will have a fifty-percent probability of observing either of the two possible outcomes; but, contrary to the mayor, my people assure me that your and Bob's photons are not entangled or correlated in any way" (and you believe the governor). Then, Bob tells you, "I measured my photon and found it to be (↑)." In this case, you conclude this new information has no bearing on your belief about your photon, so you should not change or update your predictions for future measurements.

Is it all in my head?

According to the Bayesian approach, quantum theory is not a law of Nature that physical systems 'must obey.' Rather, it is a theory that advises you about how you should decide your degree of confidence in predicting outcomes of future experiments, based on your prior knowledge and your latest observations of these systems.

The Bayesian approach seems to be a consistent way to view quantum theory, and it will never make a prediction that contradicts any other valid ways of interpreting the theory. Independent of which philosophy might be preferred by users of quantum theory, they calculate and use the same probabilities. Yet many physicists are not comfortable with the Bayesian approach when taken to its limits, wherein quantum theory represents beliefs rather than representing the physical world. One of its proponents, Ruediger Schack, acknowledges that the philosophy of **Quantum Bayesianism,** also called **QBism** for short, can be a hard pill to swallow for many physicists. He said, "When QBism holds that science is as much about the scientist as it is about the world external to the scientist, it challenges one of the most deeply held prejudices that most physicists subscribe to."[12]

Some physicists have criticized the Bayesian approach by claiming it equates a person's degree of confidence with a mere arbitrary belief. A counter to such criticism is to point out that, of course, this theory should reflect deeply Nature's ways of behaving so your degree of confidence is based on sound reasoning. This is where the detailed physics comes in, constraining what a person following the Bayesian approach will choose to believe. In other words, a careful use of the Bayesian approach will not lead one to predict one thousand fairies dancing on the head of a pin. Yet there is still a lingering awkwardness in Quantum Bayesianism, according to many physicists. Physicist Steven van Enk says, "Quantum Bayesians simply assume quantum mechanics does not describe the world, just what we know about it. But of course, then they still have to answer the question what the world is like."[13] That is, what is it about the physical world that makes it this way? Brukner writes about the problem in understanding how measurements happen:

> One possibility to address these questions would be to dismiss the measurement problem as a pseudo-issue. . . . It seems to me that this path is taken by some proponents of . . . Quantum Bayesianism (QBists), for example when Fuchs and Schack write, "a measurement is an action an agent takes to elicit an experience. The measurement outcome is the experience so elicited." Such a view is consistent and self-contained, but in my opinion, it is *not* the whole story. It is silent about the question: what makes a photon counter a better device for detecting photons than a beam splitter? Yet the question is scientifically well posed and has an unambiguous answer (which manufacturers of photo-detectors do know!).[14]

These comments take us back to our earlier discussion about decoherence and the measurement problem, and the argument that a reasonable escape from this problem is to say it reduces to a *logical* argument: IF you insist on having a permanent record of your measurement, THEN the measurement process is not necessarily reversible and therefore, by itself, is not unitary. Yet the 'jury is still out' on these questions, and other viewpoints are in competition with the Bayesian view-point, as I explain next.

Coherence forever?

Can a complex system, including even people, be described by a perfectly known quantum state with coherence between the various possibilities? Again, the typical situation in which this question arises is when a measurement is performed, as illustrated in FIGURE 14.2. Let's say a photon is known by Alice to be described perfectly by a state of diagonal polarization, which she can represent by the superposition state (↑) + (→). The photon is sent to Bob, who passes it through a calcite crystal that separates (↑) and (→) polarizations, and one of two detectors registers the outcome by generating an electron avalanche.

Detector 1 corresponds to the outcome (↑), and detector 2 corresponds to the outcome (→). Bob is observing the detector outcomes, which are indicated, say, by a light that flashes on one or the other detector. Let's assume Bob and the experiment are inside a room that lets no energy or information in or out,

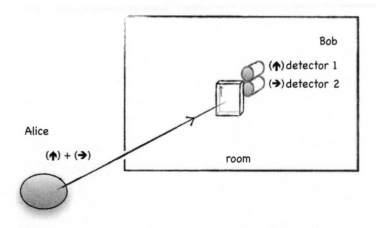

Figure 14.2 Bob, inside an 'information-proof' room, measures a photon. Alice, outside the room, writes a quantum state describing the situation of Bob and his photon.

other than the photon Alice sends. Alice is outside the room and wants to describe the goings-on inside the room using quantum theory.

After the measurement is made, which Bob sees but Alice does not, Alice might think she should describe theoretically the situation by including Bob in a quantum superposition state with the photon and the outcome, and write the state evolution as

$$\{(\uparrow) + \rightarrow)\} \,\&\, (Detector\ 1\ dark)\,\&\,(Detector\ 2\ dark)\,\&\,(Bob) \Rightarrow$$

$$(\uparrow)\,\&\,(Detector\ 1\ flashes)\,\&\,(Detector\ 2\ dark)\,\&\,(Bob\ sees\ \#1)$$
$$+$$
$$(\rightarrow)\,\&\,(Detector\ 1\ dark)\,\&\,(Detector\ 2\ flashes)\,\&\,(Bob\ sees\ \#2).$$

Is such a state physically possible? If so, it should be possible to observe quantum interference of the two possibilities in this state. Such an experiment to observe interference would almost certainly be impossible to conduct in practice. If you take the viewpoint that it is impossible to observe interference in a state describing a human, then you could argue the previous written state should not be considered a valid part of any theory.

On the other hand, many physicists take the view that such a state is fundamentally meaningful in principle because there is nothing in quantum theory itself that forbids carrying out such an interference experiment. One such view is espoused by John Preskill, Richard Feynman Professor of Theoretical Physics, California Institute of Technology.[15] Of the quantum measurement problem, Preskill writes:

> I know no good reason to disbelieve that all physical pro-cesses, including measurements, can be described by the Schrödinger equation [including superposition states]. But to describe measurement this way, we must include the observer as part of the evolving quantum system. This

[theory] does not provide us observers with deterministic predictions for the outcomes of the measurements we perform. Therefore, we are forced to use probability theory to describe these outcomes. ... The "classical" world arises due to decoherence, that is, pervasive entanglement of an observed quantum system with its unobserved environment. ... The viewpoint encapsulated [here] is a version of what is sometimes called the Everett interpretation of quantum theory. It puzzles me somewhat that physicists I respect very much [the quantum Bayesians] ... seem to find this viewpoint foolish, though perhaps I should not put words in their mouths. I admit it's less precise than one might desire, and that one can feel a bit dizzy when thinking about a description of a physical system that includes oneself.[16]

As referred to here by Preskill, the *Everett interpretation of quantum theory* was devised originally by Hugh Everett in 1957. Originally called the 'relative-state view' of quantum theory, it maintains that measurements don't yield definite outcomes, but instead only create quantum correlations, or entanglements, between the system being measured and the measuring device. Any person finding him- or herself in an entangled state along with detectors will perceive a definite outcome, but when looked at from a wider perspective, both outcome possibilities are still present in the state.

In a modern version of the Everett interpretation, taking Everett's view to its logical limit, there would be just one all-encompassing quantum state describing everything in the universe. You, me, my cat, every distant galaxy, and so on. If one assumes there is no outside observer, this quantum state is never updated. Every observer becomes entangled with his or her measurement outcomes, none of which become definite. The theory is not meant to describe particular definite outcomes, only statements such as IF such-and-such happens,

THEN thus-and-so may occur next with particular probabilities given by quantum theory. When stated this way, it doesn't sound objectionable. In fact, mathematically such an approach yields exactly the same predictions for probabilities as other ways of thinking about quantum theory. This is why such differing viewpoints are called interpretations rather than distinct theories.

This viewpoint raises again the question: Does a quantum state represent the actual world or does it represent what we know about the world? Recall Quantum Bayesians argue that the quantum state, because it represents only probabilities, which are subjective (in the mind of the person), must also be considered subjective. Preskill argues opposed to this claim, saying,

> A related controversy concerns whether the quantum state is "ontic" (a mathematical description of physical reality) or "epistemic" (a description of what a particular observer *knows* about reality). I don't really understand this question very well. Why can't there be both a fundamental ontic state for the system and observer combined, and at the same time an (arguably less fundamental) epistemic state for the system alone, which is continually updated in the light of the observer's knowledge?[17]

Such a view would seem to elevate the quantum state of the universe to an entity unto itself, which evolves over the eons according to the laws of Nature. It would not defer to human thought in any way, nor would its purpose be to serve human logic or thinking. This view of quantum theory fits well with efforts to develop a quantum theory encompassing astrophysics and cosmology—the science of the origin and development of the universe—where human thought is, presumably, not relevant.

Many researchers in astrophysics and cosmology prefer not to treat the quantum state and psi wave as mere human bookkeeping methods, but something more fundamentally physical, and as such should not be tampered with so cavalierly as they claim Heisenberg and the Quantum Bayesians do. Although they might agree that you *could* update artificially your local piece of the psi wave when gaining information from a measurement, they argue there is no need to do so, and in fact no strict justification for doing so, because there is no mechanism contained in Schrödinger's equation that tells us *how* to do so.

Then again, many physicists are skeptical about the Everett interpretation. If every observer is in a quantum entangled state with all possible outcomes of each measurement, what would that mean? One way of answering this is called the *many-worlds interpretation* of quantum theory. A leading proponent of this viewpoint is physicist David Deutsch, who writes, "Everettian quantum theory implies that generically, when an experiment is observed to have a particular result, all the other possible results also occur and are observed simultaneously by other instances of the same observer in different universes across the multiverse."[18] This would appear to mean there are many copies of you out there, some of whom had eggs for breakfast and others who did not.

An attractive aspect of this many-worlds interpretation is that, in a sense, it is simple. There is only one quantum state representing everything. There is no need to 'solve' the measurement problem. There is no need for human thought to intervene by updating the quantum state, and quantum theory can even be seen as a deterministic theory.

To many scientists, the many-worlds' unattractive aspect is that it seems absurd and cannot be tested experimentally. In their view the absurdity is the idea that when a quantum process can evolve into more than one possibility, the universe somehow coexists with many other universes—one accommodating each possibility. The other universes are not observable and are simply 'out there' somewhere, and their number is exponentially large. Many scientists dismiss this viewpoint outright, joking that "Many-worlds is cheap on assumptions, but expensive on universes."

Why do the Bell correlations occur?

I return, now, to the subject of the Bell-test experiments, which, as of 2015, have confirmed that no theory based on local causality and classical realism can describe Nature correctly. Given this now-firm conclusion, it seems shocking to many (including me) that, nevertheless, under the right experimental conditions, two distant yet quantum-entangled objects, such as photons, can yield locally random measurement outcomes that are perfectly correlated. This would be like two dancers who, separated by a great distance and not in communication, and without any preset agreement or plan, improvise identical dances at random. This conclusion (which, of course, holds for quantum objects but not actual human dancers) flies in the face of all our commonsense notions of reality. Yet experiments with photons show just this kind of correlation.

This conclusion is intriguing enough that it has stimulated deep thought about how it can be. Both the Heisenberg and the many-worlds interpretations of quantum theory are self-consistent and appeal to different groups of physicists. Spencer Chang, a young high-energy-physics theorist, sums up well the feelings of many practicing physicists:

> I do agree that causal local realism can't exist and, of course, I am a serious adherent of quantum mechanics so I believe quantum mechanics is correct. As for correlations being transmitted across large distances, I think the ways people in the field feel comfortable about it is either, i) Quantum-state "collapse" can occur across large distances, but no information is transmitted, or ii)

If one treats the measuring devices as quantum devices, measuring the polarization of each photon entangles the device with the photon—this coupled with decoherence allows us to maintain that no collapse occurred, but the many-world branches have experienced decoherence and thus on each branch the devices agree on the correct outcome. I admit that I waffle at times whether I adhere to i) or ii). Luckily, it doesn't make a practical difference in how one calculates the Bell correlations, so maybe that is why physicists have not come to a consensus.[19]

Where do the Bell correlations come from? The current mathematical form of quantum theory predicts the correct correlations with ease, but does not give insight into why or how they occur. According to quantum theory, there is no physical mechanism that causes or 'enforces' such correlations; they just happen on their own, as far as we know. There is something about the universe that is, to many scientists, deeply mysterious in this regard. Undoubtedly, the mystery is embodied in quantum state superposition and entanglement—that is, contained in the structure of quantum theory itself.

Notes

1 The Optical Society and the International Society for Optics and Photonics, "The Role of Optics and Photonics in a National Initiative in Quantum Science and Technology (QST)," February 12, 2016, http://www.lightourfuture.org/getattachment/ 66d9a226-2844-4317-b281-0e417ac39335/A-National-Initiative-in- Quantum-Technology-and-Science-110816.pdf (with permission). Note that I co-authored this report.

2 Murray Gell-Mann, *The Quark and the Jaguar: Adventures in the Simple and the Complex* (New York: W.H. Freeman, 1994), 123.

3 Academy of Achievement, "Interview: Murray Gell-Mann, Developer of the Quark Theory," December 16, 1990, http:// www.achievement.org/autodoc/printmember/gel0int-1.

4 Werner Heisenberg, *Physics and Philosophy: The Revolution in Modern Science* (New York: Harper and Row, 1962), 54.

5 Caslav Brukner, "On the Quantum Measurement Problem," in *Proceedings of the Conference "Quantum UnSpeakables II: 50 Years of Bell's Theorem,"* June 19–22, 2014, Vienna, http://arxiv.org/abs/1507.05255; quote, 4.

6 Wojciech Zurek, "Decoherence and the Transition from Quantum to Classical Revisited," *Los Alamos Science* 27 (2002), 2–25; quote, 21.

7 Brukner, "On the Quantum Measurement Problem," 10.

8 Otto Frisch, "Take a Photon," *Contemporary Physics* 7 (1965): 45–53; quote, 53.

9 Heisenberg, *Physics and Philosophy*, 55.

10 Eugene Wigner, *Symmetries and Reflections* (Bloomington, IN: Indiana University Press, 1967), 164.

11 Carlton Caves, Christopher Fuchs, and Ruediger Schack, "Subjective Probability and Quantum Certainty," *Studies in History and Philosophy of Science Part B: Studies in History and Philosophy of Modern Physics* 38 (2007): 255–274; quote, 255.

12 Ruediger Schack, interview by Luke Muehlhauser for the Machine Intelligence Research Institute, April 29, 2014, https://intelligence.org/2014/04/29/ruediger-schack/.

13 Steven van Enk, private communication, 2015.

14 Brukner, "On the Quantum Measurement Problem," 4.

15 Preskill has made many contributions to theoretical physics, but his claim to popular media fame is that he won a bet with Steven Hawking on the fate of quantum information inside black holes.

16 John Preskill, "A Poll on the Foundations of Quantum Theory," in *Quantum Frontiers*, January 10, 2013, http://quantumfrontiers.com/2013/01/10/a-poll-on-the-foundations-of-quantum-theory/.

17 Preskill, "A Poll on the Foundations of Quantum Theory."

18 David Deutsch, "The Logic of Experimental Tests, Particularly of Everettian Quantum Theory," 2015, http://arxiv.org/abs/1508.02048. This piece was later published with revisions in *Studies in History and Philosophy of Modern Physics* 55 (2016): 24–33.

19 Spencer Chang, private communication, 2015.

INDEX

Note: figures are indicated by an italic *f* following the page number.